A. Ghosh, L.C. Jain (Eds.)

Evolutionary Computation in Data Mining

To our students

A. Ghosh
L. C. Jain

T0189778

Studies in Fuzziness and Soft Computing, Volume 163

Editor-in-chief
Prof. Janusz Kacprzyk
Systems Research Institute
Polish Academy of Sciences
ul. Newelska 6
01-447 Warsaw
Poland
E-mail: kacprzyk@ibspan.waw.pl

Further volumes of this series
can be found on our homepage:
springeronline.com

Ashish Ghosh
Lakhmi C. Jain (Eds.)

Evolutionary Computation in Data Mining

 Springer

Dr. Ashish Ghosh
Indian Statistical Institute
Machine Intelligence Unit
203 Barrackpore Trunk Road
Kolkata 700 108
India
E-mail: ash@isical.ac.in

Prof. Lakhmi C. Jain
University of South Australia
Knowledge-Based Intelligent
Engineering Systems Centre
Mawson Lakes
5095 Adelaide
Australia
E-mail: l.jain@unisa.edu.au

ISSN 1434-9922
ISBN 3-642-42195-4   Springer Berlin Heidelberg New York

Springer is a part of Springer Science+Business Media
springeronline.com

Typesetting: data delivered by editors
Cover design: E. Kirchner, Springer-Verlag, Heidelberg
Printed on acid free paper   62/3020/M - 5 4 3 2 1 0

Preface

Data mining (DM) consists of extracting interesting knowledge from real-world, large & complex data sets; and is the core step of a broader process, called the knowledge discovery from databases (KDD) process. In addition to the DM step, which actually extracts knowledge from data, the KDD process includes several preprocessing (or data preparation) and post-processing (or knowledge refinement) steps. The goal of data preprocessing methods is to transform the data to facilitate the application of a (or several) given DM algorithm(s), whereas the goal of knowledge refinement methods is to validate and refine discovered knowledge. Ideally, discovered knowledge should be not only accurate, but also comprehensible and interesting to the user. The total process is highly computation intensive.

The idea of automatically discovering knowledge from databases is a very attractive and challenging task, both for academia and for industry. Hence, there has been a growing interest in data mining in several AI-related areas, including evolutionary algorithms (EAs). The main motivation for applying EAs to KDD tasks is that they are robust and adaptive search methods, which perform a global search in the space of candidate solutions (for instance, rules or another form of knowledge representation).

The evolutionary computing community has been publishing KDD-related articles in a relatively scattered manner in conference proceedings/journals dedicated to knowledge discovery and data mining or evolutionary computing. The objective of this volume is to assemble a set of high-quality original contributions that reflect and advance the state-of-the-art in the area of Data Mining and Knowledge Discovery with Evolutionary Algorithms. The book will also emphasize the utility of different evolutionary computing tools to various facets of KDD, ranging from theoretical analysis to real-life applications.

This topic is quite new and there are only a few books in the literature. This book discusses advanced theories of evolutionary computing/data mining, and recent applications like web mining or bioinformatics. The book contains twelve chapters written by leading experts of the field demonstrating how different evolutionary computing tools can be used for solving real life problems in data mining, web mining, bioinformatics in addition to provide fundamentals of evolutionary computing and data mining. This provides a balance mixture of theory, algorithms and applications in cohesive manner.

The book starts with an introductory chapter by one of the editors (Ashish Ghosh) where he discusses the use of evolutionary algorithms, particularly genetic algorithms and genetic programming, in data mining and knowledge discovery with some applications.

Chapters 2-4 describe how to use EC for feature selection/preprocessing. In Chapter 2, Cano, Herrera and Lozano have carried out an empirical study using different size data sets to evaluate the scaling up problem. The results show that the stratified evolutionary instance selection algorithms consistently outperform the non-evolutionary ones. The main advantages that they found are: better instance reduction rates, higher classification accuracy and reduction in resources consumption. In the next chapter Smith and Bull have examined the use of Genetic Programming and a Genetic Algorithm to pre-process data before it is classified using the C4.5 and decision tree learning algorithms. Genetic Programming is used to construct new features from those available in the data, a potentially significant process for data mining since it gives consideration to hidden relationships between features. A Genetic Algorithm is used to determine which such features are the most predictive. Using ten well-known datasets they have shown that their approach, in comparison to C4.5 alone, provided marked improvement in a number of cases. In the fourth chapter Sikora presented a multi-agent based inductive learning algorithm for scaling up and improving the performance of traditional algorithms; the system used genetic algorithms as learning agents incorporating a self-adaptive feature selection method.

Chapters 5-7 discuss the use of EC for rule generation. In Chapter 5 Fu and Wang proposed a decompositional rule extraction method based on radial basis function neural networks. In the proposed rule extraction method, rules are extracted from trained RBF networks with class-dependent features. Genetic Algorithm is used to determine the feature subsets corresponding to different classes. Rules are extracted from trained RBF networks by a gradient descent method. The chapter by Yu, Tan and Lee discusses a coevolution-based classification technique, which they call CORE (COevolutionary Rule Extractor), that discovers cohesive classification rules. The proposed system coevolves rules and rule sets concurrently in two cooperative populations to confine the search space and to produce good rule sets that are cohesive and comprehensive. Comparison results show that the proposed CORE produces comprehensive and good classification rules for most datasets, obtained from UCI machine learning repository, which are competitive as compared with existing classifiers in literature. Nguyen, Abbass, and McKay showed in Chapter 7 that combining different neural networks can improve the generalization ability of learning machines for rule discovery. Diversity of the ensemble's members plays a key role in minimizing the combined bias and variance of the ensemble. In this chapter, we compare between different mechanisms and methods for promoting diversity in an ensemble. In general, they found that it is important to design the diversity promoting mechanism very carefully for the ensemble's performance to be satisfactory.

In Chapter 8 Nasraoi and Krishnapuram prsented a *robust* clustering algorithm, called the *Unsupervised Niche Clustering algorithm (UNC)*, that overcomes all the above difficulties. UNC can successfully find dense areas (clusters) in feature space and determines the *number* of clusters *automatically*. Robust cluster scale estimates were *dynamically* estimated using a hybrid learning scheme coupled with the genetic optimization of the cluster centers, to adapt to clusters of different sizes and noise contamination rates. Genetic Optimization enables this approach to handle data with both numeric and qualitative attributes, and general *subjective, non metric, even non-differentiable* dissimilarity measures.

Next four chapters are on application of EC in several aspects of data mining. In Chapter 9 Abraham showed how EC can be used for intrusion detection in computer systems, as well as web usage mining to discover useful knowledge from the secondary data obtained from the interactions of the users with the web. He also compared the performance with other techniques to establish the superiority of EC based techniques. In Chapter 10 Langdon and Barrett used genetic programming (GP) to automatically create interpretable predictive models of oral bioavailability of a small number of complex biological interactions that are of great interest to medicinal and computational chemists who search for new drug treatments. The models can make *in silico* predictions about "virtual" chemicals, e.g. to decide if they are to be synthesized. FOGEL, in Chapter 11, provided a recent state-of-the-art literature survey on the application of evolutionary algorithms in the area of microarray data establishing that simulated evolution provides better predictive models for the same. In Chapter 12 Ko discussed a modularized financial distress forecasting mechanism based on evolutionary algorithm, which allows using any evolutionary algorithm, such as Particle Swarm Optimization, Genetic Algorithm and etc., to extract the essential financial patterns. One more evaluation function modules, such as Logistic Regression, Discriminant Analysis, Neural Network, are integrated to obtain better forecasting accuracy by assigning distinct weights, respectively. Specifically they applied evolutionary algorithm to select critical financial ratios and obtained better forecasting accuracy. This model when integrated by other classification models like logistic regression, neural networks etc. showed a much better performance.

We are grateful to the contributors and the referees for their vision and efforts.

Kolkata, India, April 2004   *Ashish Ghosh*
Adelaide, Australia, April 2004   *Lakhmi C. Jain*

Table of Contents

List of Contributors

A. Abraham
Natural Computation Lab
Department of Computer Science
Oklahoma State University,USA
ajith.abraham@ieee.org

H. Abbass
Artificial Life and Adaptive Robotics
(A.L.A.R.) Lab
School of Information Technology
and Electrical Engineering
Australian Defence Force Academy
University of New South Wales,
Canberra, ACT 2600, Australia
h.abbass@adfa.edu.au

S. J. Barrett
Data Exploration Sciences
GlaxoSmithKline
Research and Development
Greenford, Middlesex
UK

L. Bull
Faculty of Computing, Engineering
& Mathematical Sciences
University of the West of England
Bristol BS16 1QY
U.K.
Larry.Bull@uwe.ac.uk

J. R. Cano
Dept. of Electronic Engineering
Computer Systems and Automatics
Escuela Superior de La Rabida
University of Huelva
21819, Huelva, Spain
jose.cano@diesia.uhu.es

G. B. Fogel
Natural Selection, Inc.
3333 N. Torrey Pines Ct., Suite 200
La Jolla, California 92037
USA
gfogel@natural-selection.com

X. J. Fu
Institute of High Performance Computing
Science Park 2, 117528
Singapore
fuxj@ihpc.a-star.edu.sg

A. Ghosh
Machine Intelligence Unit
Indian Statistical Institute
203 B. T. Road
Kolkata 700 108, India
ash@isical.ac.in

F. Herrera
Dept. of Computer Science and Artificial Intelligence
University of Granada
18071, Granada, Spain
herrera@decsai.ugr.es

P. C. Ko
National Kaohsiung University of
Applied Sciences

Department of Information Management Kaohsiung
Taiwan 807, R.O.C
cobol@cc.kuas.edu.tw

W. B. Langdon
Data Exploration Sciences
GlaxoSmithKline,
Research and Development,
Greenford, Middlesex,
UK

T. H. Lee
Department of Electrical and Computer Engineering
National University of Singapore
4 Engineering Drive 3
Singapore 117576

Elizabeth Leon
Department of Electrical and Computer Engineering
The University of Memphis,
206 Engineering Science Bldg.
Memphis, TN 38152-3180

P. C. Lin
Van Nung Institute of Technology
Department of Information Management
Jungli, Taiwan 320, R.O.C
lety@cc.vit.edu.tw

M. Lozano
Dept. of Computer Science and Artificial Intelligence
University of Granada
18071, Granada, Spain
lozano@decsai.ugr.es

R. McKay
Artificial Life and Adaptive Robotics (A.L.A.R.) Lab
School of Information Technology

and Electrical Engineering
Australian Defence Force Academy
University of New South Wales,
Canberra, ACT 2600, Australia

Olfa Nasraoui
Department of Electrical and Computer Engineering
The University of Memphis
206 Engineering Science Bldg.
Memphis, TN 38152-3180
onasraou@memphis.edu

Raghu Krishnapuram
IBM India Research Lab
Block 1, Indian Institute of Technology
Hauz Khas, New Delhi 110016, India
kraghura@in.ibm.com

M. H. Nguyen Artificial Life and Adaptive Robotics (A.L.A.R.) Lab
School of Information Technology
and Electrical Engineering
Australian Defence Force Academy
University of New South Wales,
Canberra, ACT 2600, Australia
m.nguyen@adfa.edu.au

R. Sikora
Dept. of Information Systems & OM
The University of Texas at Arlington
P.O. Box 19437, Arlington, TX 76019
rsikora@uta.edu

M. G. Smith
Faculty of Computing, Engineering & Mathematical Sciences
University of the West of England
Bristol BS16 1QY
U.K.
Matt-Smith@bigfoot.com

K. C. Tan
Department of Electrical and Computer Engineering
National University of Singapore
4 Engineering Drive 3
Singapore 117576
eletankc@nus.edu.sg

L. P. Wang
School of Electrical and Electronic Engineering
Nanyang Technological University (NTU)
Nanyang Avenue, Singapore 639798
elpwang@ntu.edu.sg

Q. Yu
Department of Electrical and Computer Engineering
National University of Singapore
4 Engineering Drive 3
Singapore 117576

1. Evolutionary Algorithms for Data Mining and Knowledge Discovery

Ashish Ghosh

Machine Intelligence Unit
Indian Statistical Institute
203 B. T. Road
Kolkata 700 108, India
ash@isical.ac.in

Abstract: This chapter discusses the use of evolutionary algorithms, particularly genetic algorithms, in data mining and knowledge discovery. We focus on the data mining task of rule generation. We show how the requirements of data mining and knowledge discovery influence the design of evolutionary algorithms. In particular, we discuss how individual representation, genetic operators and fitness functions have to be adapted for extracting rules from data. We also discussed an algorithm for using multi-objective evolutionary algorithms for rule mining.

1.1 Introduction

Since 1960, database technology and information technology have been evolving systematically from primitive file processing to sophisticated powerful database systems through the development of DBMS. But the explosive growth in the data collected from various application areas are continuously increasing in recent times. Intuitively, this large amount of stored data contains valuable hidden knowledge, which could be used to improve the decision-making process of an organization. For instance, data about previous sales might contain interesting relationships between products and customers. The discovery of such relationships can be very useful to increase the sale of a company. No human can use such a big volume of data in an efficient way. Thus, there is a clear need for semi-automatic methods for extracting knowledge from data. This need has led to the emergence of a field called data mining and knowledge discovery (KDD) [1.16, 1.5, 1.1, 1.27]. This is an interdisciplinary field, using methods of several research areas (specially machine learning and statistics) to extract knowledge from real-world data sets. Data mining is the core step of a broader process, called knowledge discovery from databases. This whole process includes application of several preprocessing methods aimed at facilitating application of the data mining algorithms and postprocessing methods for refining and improving the discovered knowledge.

This chapter discusses the use of evolutionary algorithms (EAs), particularly genetic algorithms (GAs) [1.15, 1.22] in KDD process. We mainly focus on rule mining. We show how the requirements of data mining and knowledge discovery influence the design of GAs. In particular, we discuss how individual representation, genetic operators and fitness functions have to be adapted for extracting knowledge from data bases. We also discussed an algorithm for using multi-objective evolutionary algorithms for rule mining.

This chapter is organized as follows. Section 2 presents an overview of data mining and knowledge discovery process. Section 3 discusses several aspects of the design of GAs for rule discovery. Section 4 discusses the use of multi-objective GAs for rule discovery. Finally, Section 5 presents a discussion that concludes the chapter.

1.2 Knowledge Discovery in Databases

Data collected from various application domains is becoming increasingly high in recent time. They are stored in various repositories like different databases (such as relational databases, transactional databases, object-relational databases, object-oriented databases etc.), different data warehouses (repository of multiple heterogeneous data for long term and providing online analytical processing), and other repositories (such as WWW data, biological data, text data, image data, multimedia data etc.).

1.2.1 Data Preprocessing

During the recording or storing of data in the repositories, it is obvious that some error will be there. Such errors may be missing values, noise, inconsistency etc and the data may not be in suitable format for processing. So, before going to perform mining on the data some kind of preprocessing [1.16, 1.18, 1.24] is required. Preprocessing of data is done in the following major ways.

Data cleaning. To remove inconsistency, noise , to fill up missing values, to identify the outliers etc., data cleaning is performed. The missing values are filled up manually or using some global constants. Noises are smoothed using binning, regression analysis etc.

Data integration. This is needed to combine data from multiple sources like databases, data cubes, flat files etc. The integration can be done in terms of metadata, correlation analysis, detecting data conflict, and resolving semantic heterogeneity. Correlation analysis measures how one attribute strongly implies the other based on available data. Data conflicts are detected by identifying the difference among the real world attribute values [say price can be expressed in Rs, $ etc.]. Finally, the semantic heterogeneities are removed.

Data transformation. The format of data in the repositories may not be suitable for processing. So, the format of the data should be transformed to a one suitable for a particular task. This is done for smoothing, aggregation, generalization, normalization, and attribute construction. Smoothing uses binning, clustering, regression analysis etc. Aggregation is done in terms of summarization [say daily sales may be summarized or aggregated to compute monthly, yearly etc.]. Generalization is done using the concept of "hierarchies of data" where low-level data are replaced by high-level data [say, for age we can represent as young, old, senior, junior etc]. Normalization is done in terms of scaling the attribute data so as to fall within a small range [say, to have data values between the range 0.0 to 1.0]. New attributes are constructed from the given data sets with desired format.

Discretization. This step consists of transforming a continuous attribute into a categorical (or nominal) attribute, taking only a few discrete values - e.g., the real-valued attribute. Salary can be discretized to take on only three values, say "low", "medium", and "high". This step is particularly required when the data mining algorithm cannot cope with continuous attributes. In addition, discretization often improves the comprehensibility of the discovered knowledge.

Data reduction. To have reduced representation of data sets that is necessary for efficient mining this step is used. This data reduction is done in terms of dimensionality reduction, data cube aggregation, data compression, numerosity reduction etc. Dimensionality reduction is done removing some features from the data sets. Different data cubes are aggregated in order to have desired representation of data. Numerosity of the data is reduced using alternatives like sampling, histogram etc.

Data selection. For the purpose of processing and analysis, relevant data are selected and retrieved in this step.

1.2.2 Data Mining

In order to extract or mine the knowledge or pattern of interest from the data that is ready after preprocessing task, intelligent mining tools are applied. These tools or methods are *association analysis, clustering, classification and prediction, dependence analysis,* and *outlier analysis.*

Association analysis:. Discovery of association relationship among large set of data items is useful in decision-making. This can be better understood with a typical example called *market basket analysis*, which studies customer buying habits by finding associations between the different items that customers place in their baskets. For instance, if customers are buying milk, how likely are they to buy bread on the same trip to the market? An association rule is thus a relationship of the form IF A THEN B, where A and B are sets of items and $A \cap B = \emptyset$. Such a rule generation technique consists of finding frequent item sets (set of items, such as A and B satisfying minimum support and minimum confidence) from which rules like A=>B are generated. The measures support is the percentage of transactions that contain both the item sets. Thus Support (A U B) = p(A U B)

Confidence (A=>B) $= \frac{p(A \cup B)}{p(A)}$

Classification and prediction. Classification is a process of finding a set of models or functions that describes and distinguishes data classes, for the purpose of using it. This derived model is based on the analysis of a set of training data object (i.e. data objects whose class labels are known).

Classification rules can be considered as a particular kind of prediction rules where the rule antecedent ("IF part") contains a combination - typically, a conjunction - of conditions on predicting attributes, and the rule consequent ("THEN part") contains a predicted value for the goal attribute. Examples of classification rules are: IF (Unpaid_Loan? = "no") and (Current_account_balance > $ 3,000) THEN (Credit = "good") IF (Unpaid_Loan = "yes") THEN (Credit = "bad").

Although both classification and association rules have an IF-THEN structure, there are major differences between them. Major differences are: association rules can have more than one item in the consequent part, whereas classification rules always have one attribute (the goal one) in the consequent. In other words, for classification rules , predicting attributes and the goal attribute. Predicting attributes can occur only in the rule antecedent, whereas the goal attribute occurs only in the rule consequent.

Dependence modelling. This task can be regarded as a generalization of the classification task. In the former we want to predict the value of several attributes - rather than a single goal attribute, as in classification. In its most general form, any attribute can occur both in the antecedent of a rule and in the consequent of another rule - but not in both the antecedent and the consequent of the same rule. For instance, we might discover the following two rules:

IF (Current_account_balance > $3,000) AND (Salary = "high")
THEN (Credit = "good")
IF (Credit = "good") AND (Age > 21) THEN (Grant_Loan = "yes").

Clustering. In a huge data set objects can be identified to keep them in different groups. The process of grouping a set of data objects that are similar to one another within the same cluster and are dissimilar to objects in other cluster is called clustering. The similarity and dissimilarity are measured in terms of some distance measure.

For classification the class label of a training example is given as input to the classification algorithm, characterizing a form of supervised learning. In contrast, the clustering task discover classes by itself, by partitioning the examples into clusters, which is a form of unsupervised learning [1.8]. Note that, once the clusters are found, each cluster can be considered as a "class", and one can run a classification algorithm on the clustered data, by using the cluster name as a class label.

So many clustering algorithms have been developed: and they can broadly be classified as partitioning method, hierarchical method, density-based method, grid based method, model based method.

Outlier analysis:. In the whole data set, some data objects do not comply with the general behavior of the data. They are grossly different or inconsistent from the remaining set of data. These data objects are called outliers. For example, salary of the chief executive to other employees in a company can be considered as an outlier for the salary data set. This kind of outliers can be analyzed in terms of detection of these outlier data objects using various approaches including *statistical, distance-based* and *deviation-based* approaches are used.

1.2.3 Knowledge Interpretation or Postprocessing

Knowledge interpretation is the last step in the KDD process. Extracted patterns are required to be interpreted properly, so that they can be used for decision making; i.e. to interpret the knowledge that the patterns are carrying. Patterns representing knowledge are evaluated i.e. identified properly by interestingness measures.

A pattern will be interesting if

- it is easily understood by humans (i.e. simplicity of the pattern)
- valid on test data with some degree of certainty
- potentially useful
- novel,
- validates a hypothesis that the user sought to confirm.

Once the patterns are evaluated and discovered, they are presented to the users for interaction and to guide them for further discovery. This is done using *visualization* techniques, which includes tables, crosstabs (cross-tabulations), pie charts, bar charts, decision trees, rules etc.

It is often the case that the knowledge discovered by a data mining algorithm needs to undergo some kind of postprocessing. There are two main motivations for such postprocessing. First, when the discovered rule set is large, we often want to simplify it - i.e. to remove some rules and/or rule conditions - in order to improve knowledge comprehensibility for the user. Second, we often want to extract a subset of interesting rules, among the discovered ones. The reason is that although many data mining algorithms were designed to discover accurate, comprehensible rules, most of these algorithms were not designed to discover interesting rules, which is a rather more difficult and ambitious goal.

Methods for selection of interesting rules can be roughly divided into subjective and objective methods. Subjective methods are user-driven and domain-dependent. For instance, the user may specify rule templates, indicating which combination of attributes must occur in the rule for it to be considered interesting - this approach has been used mainly in the context of association rules [1.11]. Another example of a subjective method, the user can give the system a general, high-level description of his/her previous knowledge about the domain, so that the system can select only the discovered rules which represent previously-unknown knowledge for the user [1.11].

By contrast, objective methods are data-driven and domain-independent. Some of these methods are based on the idea of comparing a discovered rule against other rules, rather than against the user's beliefs. In this case the basic idea is that the interestingness of a rule depends not only on the quality of the rule itself, but also on its similarity to other rules. [1.11].

1.2.4 Desirable Properties of Discovered Knowledge

In essence, data mining consists of the semi-automatic extraction of knowledge from data. This statement raises the question of what kind of knowledge we should try to discover. Although this is a subjective issue, we can mention three general properties that the discovered knowledge should satisfy; namely, it should be accurate, comprehensible, and interesting. Let us briefly discuss each of these properties in turn.

As will be seen in the next subsection, in data mining we are often interested in discovering knowledge which has a certain predictive power. The basic idea is to predict the value that some attribute(s) will take on in "future", based on previously observed data. In this context, we want the discovered knowledge to have a high predictive accuracy.

We also want the discovered knowledge to be comprehensible for the user. This is necessary whenever the discovered knowledge is to be used for supporting a decision to be made by a human being. If the discovered "knowledge" is just a black box, which makes predictions without explaining them, the user may not trust it [1.16]. Knowledge comprehensibility can be achieved by using high-level knowledge representation. A popular one is a set of IF-THEN rules, where each rule is of the form:

IF <some_conditions_are_satisfied> THEN
<predict_some_value_for_an_attribute>.

The third property, knowledge interestingness, is the most difficult one to define and quantify, since it is, to a large extent, subjective. However, there are some aspects of knowledge interestingness that can be defined in objective terms.

1.2.5 Applications

Spatial data mining. A spatial database stores a large amount of space-related data, such as maps, preprocessed remote sensing or medical imaging data and VLSI chip layout data. They carry topological and/or distance information and usually organized by multidimensional structure using data cubes. Spatial data mining refers to the extraction of knowledge like spatial relationship or other interesting patterns from large geo-spatial databases.

Web mining. With the explosive growth of information sources available on the World Wide Web (WWW), it has become increasingly necessary for users to utilize automated tools in order to find, filter, and evaluate the desired information and resources [1.6, 1.23]. Web mining can be broadly defined as the discovery and analysis of useful information from the WWW. In order to mine the web basically two ideas are used.

Web content mining: here the idea is automatic search and retrieval of the information.

Web usage mining: the basic idea here is to automatic discovery and analysis of user access patterns from one or more web servers.

Text mining. In recent days we can have databases, which contain large collection of documents from various sources such as news articles, research papers, books, digital libraries, e-mail messages, and various web pages which are called text databases or document databases. These text databases are rapidly growing due to the increasing amount of information available in electronic forms, such as electronic publications, e-mails, CD-ROMs etc.

Data stored in most text databases are semi-structured data, in that they are neither completely unstructured nor completely structured. For example, a document may contain a few structured fields, such as title, authors, publication date, category and so on, but also contain some largely unstructured text component such as abstract and contents. This type of text data has challenges to the traditional retrieval techniques. As a result, text-mining concepts are increasingly coming into light. Text mining goes one step beyond the traditional approach and discovers knowledge from semi-structured text data also.

Image mining. Actually image mining i.e. mining the image databases falls under the multimedia database mining, which also contains audio data, video data along with image data. Basically images are stored with some description against a particular image. Again images are nothing but some intensity values, which figure the image in terms of color, shape, texture etc. The mining task is based on using such information contained in the images. Based on this image mining techniques can be categorized in two places:

- description based retrieval, and
- content based retrieval

Biological data mining. Biological researches are dealing greatly in development of new pharmaceutical, various therapies, medicines and human genome by discovering large-scale sequencing patterns and gene functions. In the process of gene technology the DNA data analysis becomes significantly focused with data mining applications, since the discovery of genetic causes for many diseases and disabilities and to discover new medicines as well as disease diagnosis, prevention, and treatment DNA analysis is a must.

The DNA sequences form the foundation of the genetic code of all living organisms. All DNA sequences comprised four basic nucleotides [i.e. Adenine (A), Cytosine (C), Guanine (G), Thiamine (T)]. These four nucleotides are combined in different orders to form long sequences or chains in the structure of DNA.

There are almost an unlimited number of ways that the nucleotides can be ordered and sequenced which play important role in various diseases. It is a challenging task to identify such a particular sequence from among the unlimited sequences, which are actually responsible for various diseases. Now people are trying to use data mining techniques to search and analyze these sequence patterns.

In addition to DNA sequencing, linkage analysis, and association analysis (where the structure, function, next generation genes, co-occurring genes etc) are also studied.

For all these machine learning, association analysis, pattern matching, sequence alignments, Bayesian learning etc techniques are being used in bio-informatics recently [1.17, 1.26].

A recent technology called *Microarry*, is being used in bio-informatics, which provides the experimental approach to measure levels of gene expression, subject to growth, stress and some other conditions. Levels are determined by some ratio of signal expression with the help of laser scanner. Then the intensity values are used to create an image file, from where image-mining techniques are invoked

Scientific data mining. Computational simulations and data acquisition in a variety of scientific and engineering domains have made tremendous progress over the past few decades [1.25]. Coupled with the availability of massive storage systems and fast networking technology to manage and assimilate data, these have given a significant impetus to data mining in scientific domains. In other sense, we can say that data mining has become a key computational technology applying in various domains such as geological and geophysical applications, astrophysics, bio-informatics, chemical sciences etc.

Distributed Data Mining. The evolution of KDD system from being centralized and stand alone along the dimension of data distribution signifies the emergence of distributed data mining (DDM). More precisely in that, when data mining is undertaken in an environment, where users, data, hardware, and the mining software are geographically dispersed, will be called DDM. Typically such environments are also characterized by the heterogeneity of data, multiple users, and large data volumes.

In the following sections, we describe the utility of GAs for a specific problem of data mining, namely rule mining.

1.3 Genetic Algorithms (GAs) for Rule Discovery

1.3.1 Association Rule Mining Algorithms: An Overview

Existing algorithms for mining association rules are mainly based on the approach suggested by Agrawal et al. [1.2]. Apriori [1.2], SETM [1.9], AIS [1.2], Pincer search [1.10], DIC [1.4] etc. are some of the popular algorithms based on this approach. These algorithms work on a binary database, termed as *market basket database*. On preparing the market basket database, every record of the original database is represented as a binary record where the fields are defined by a unique value of each attribute in the original database. The fields of this binary database are often termed as an *item*. For a database having a huge number of attributes and each attribute containing a lot of distinct values, the total number of items will be huge. Storing of this binary database, to be used by the rule mining algorithms, is one of the limitations of the existing algorithms.

Another aspect of these algorithms is that they work in two phases. The first phase is for *frequent item-set* generation. Frequent item-sets are detected from all-possible item-sets by using a measure called *support count* (SUP) and a user-defined parameter called *minimum support*. Support count of an item set is defined by the number of records in the database that contains all the items of that set. If the value of *minimum support* is too high, number of frequent item sets generated will be less, and thereby resulting in generation of few rules. Again, if the value is too small, then almost all possible item sets will become frequent and thus a huge number of rules may be generated. Selecting better rules from them may be another problem.

After detecting the frequent item-sets in the first phase, the second phase generates the rules using another user-defined parameter called *minimum confidence* (which again affects the generation of rules). Confidence factor or predictive accuracy of a rule is defined as

Confidence = SUP (AUC)/SUP (A) for a rule $A \rightarrow C$.

Another limitation of these algorithms is the encoding scheme where separate symbols are used for each possible value of an attribute. This encoding scheme may be suitable for encoding the categorical valued attributes, but not for encoding the numerical valued attributes as they may have different values in every record. To avoid this situation, some ranges of values may be defined. For each range of values an item is defined. This approach is also not suitable for all situations. Defining the ranges will create yet another problem, as the range of different attributes may be different.

GA based approach is free from this limitations.

Existing algorithms, try to measure the quality of generated rules by considering only one evaluation criterion, i.e., *confidence factor* or *predictive accuracy*. This criterion evaluates the rule depending on the number of occurrence of the rule in the entire database. More the number of occurrences better is the rule. The generated rule may have a large number of attributes involved in the rule thereby making it difficult to understand [1.11]. If the generated rules are not understandable to the user, the user will never use them. Again, since more importance is given to those rules, satisfying number of records, these algorithms may extract some rules from the data that can be easily predicted by the user. It would have been better for the user, if the algorithms can generate some of those rules that are actually hidden inside the data. These algorithms do not give any importance towards the rare events, i.e., interesting rules [1.11].

In the present work we used the *comprehensibility* and the *interestingness* measure of the rules in addition to predictive accuracy. In the next section, we will discuss about them in detail. Using these three measures - comprehensibility, interestingness and the predictive accuracy, some previously unknown, easily understandable rules can be generated. Hence, the rule-mining problem is not a single objective problem; rather we visualize them as a multi-objective problem.

In general the main motivation for using GAs in the discovery of high-level prediction rules is that they perform a global search and cope better with attribute interaction than the greedy rule induction algorithms often used in data mining.

1.3.2 Individual Representation

GAs for rule discovery can be divided into two broad approaches, based on how rules are encoded in the population of individuals ("chromosomes"). In the Michigan approach each individual encodes a single prediction rule, whereas in the Pittsburgh approach each individual encodes a set of prediction rules.

The choice between these two approaches strongly depends on which kind of rule we want to discover. This is related to which kind of data mining task we are addressing. Suppose the task is classification. Then we usually evaluate the quality of the rule set as a whole, rather than the quality of a single rule. In other words, the interaction among the rules is important. In this case, the Pittsburgh approach seems more natural.

On the other hand, the Michigan approach might be more natural in other kinds of data mining tasks. An example is a task where the goal is to find a small set of high-quality prediction rules, and each rule is often evaluated independently of other rules.

The Pittsburgh approach directly takes into account rule interaction when computing the fitness function of an individual. However, this approach leads to syntactically-longer individuals, which tends to make fitness computation more computationally expensive. In addition, it may require some modifications to standard genetic operators to cope with relatively complex individuals.

By contrast, in the Michigan approach the individuals are simpler and syntactically shorter. This tends to reduce the time taken to compute the fitness function and to simplify the design of genetic operators. However, this advantage comes with a cost. First of all, since the fitness function evaluates the quality of each rule separately, now it is not easy to compute the quality of the rule set as a whole - i.e. taking rule interactions into account. Another problem is that, since we want to discover a set of rules, rather than a single rule, we cannot allow the GA population to converge to a single individual - which is what usually happens in standard GAs. This introduces the need for some kind of niching method [1.21], which obviously is not necessary in the case of the Pittsburgh approach. We can avoid the need for niching in the Michigan approach by running the GA several times, each time discovering a different rule. The drawback of this approach is that it tends to be computationally expensive.

1.3.3 Representing the Rule Antecedent (a Conjunction of Conditions)

A simple approach to encode rule conditions into an individual is to use a binary encoding. Suppose that a given attribute can take on k discrete values. Then we can encode a condition on the value of this attribute by using k bits. The i-th value (i=1,...,k) of the attribute domain is part of the rule condition if and only if the i-th bit is "on" [1.11].

For instance, suppose that a given individual represents a rule antecedent with a single attribute-value condition, where the attribute is Marital_Status and its values can be "single", "married", "divorced" and "widow". Then a condition involving this attribute would be encoded in the genome by four bits. If these bits take on, say, the values "0 1 1 0" then they would be representing the following rule antecedent: IF (Marital_Status = "married" OR "divorced").

Hence, this encoding scheme allows the representation of conditions with internal disjunctions, i.e. with the logical OR operator within a condition.

Obviously, this encoding scheme can be easily extended to represent rule antecedents with several conditions (linked by a logical AND) by including in the genome an appropriate number of bits to represent each attribute-value condition.

Note that if all the k bits of a given rule condition are "on", this means that the corresponding attribute is effectively being ignored by the rule antecedent, since any value of the attribute satisfies the corresponding rule condition. In practice, it is desirable to favor rules where some conditions are "turned off" - i.e. have all their bits set to "1" – in order to reduce the size of the rule antecedent. (Recall that we want comprehensible rules and, in general, the shorter the rule is the more comprehensible it is.) To achieve this, one can automatically set all bits of a condition to "1" whenever more than half of those bits are currently set to "1".

The above discussion assumed that the attributes were categorical, also called nominal or discrete. In the case of continuous attributes the binary encoding mechanism gets slightly more complex. A common approach is to use bits to represent the value of a continuous attribute in binary notation. For instance, the binary string "0 0 0 0 1 1 0 1" represents the value 13 of a given integer-valued attribute.

Instead of using a binary representation for the genome of an individual, this genome can be expressed in a higher-level representation which directly encodes the rule conditions. One of the advantages of this representation is that it leads to a more uniform treatment of categorical and continuous attributes, in comparison with the binary representation.

In any case, in rule discovery we usually need to use variable-length individuals, since, in principle, we do not know a priori how many conditions will be necessary to produce a good rule. Therefore, we might have to modify crossover to be able to cope with variable-length individuals in such a way that only valid individuals are produced by this operator.

1.3.4 Representing the Rule Consequent (Predicted Class)

Broadly speaking, there are three ways of representing the predicted class (the "THEN" part of the rule) in an evolutionary algorithm. The first possibility is to encode it in the genome of an individual – possibly making it subject to evolution.

The second possibility is to associate all individuals of the population with the same predicted class, which is never modified during the running of the algorithm. Hence, if we want to discover a set of classification rules predicting k different classes, we would need to run the evolutionary algorithm at least k times, so that in the i-th run, i=1,..,k, the algorithm discovers only rules predicting the i-th class [1.19].

The third possibility is to choose the predicted class most suitable for a rule, in a kind of deterministic way, as soon as the corresponding rule antecedent is formed. The chosen predicted class can be the class that has more representatives in the set of examples satisfying the rule antecedent [1.13] or the class that maximizes the individual's fitness [1.11].

The first and third possibilities have the advantage of allowing different individuals of the population to represent rules predicting different classes. This avoids the need to perform multiple runs of the evolutionary algorithm to discover rules predicting different classes, which is the case in the second possibility. Overall, the third possibility seems more sound.

1.3.5 Genetic Operators for Rule Discovery

There has been several proposals of genetic operators designed particularly for rule discovery. We review some of these operators in the following subsections.

Selection. REGAL [1.14] follows the Michigan approach, where each individual represents a single rule. Since the goal of the algorithm is to discover a set of (rather than just one) classification rules, it is necessary to avoid the convergence of the population to a single individual (rule).

REGAL does that by using a selection procedure called universal suffrage. In essence, individuals to be mated are "elected" by training examples. An example "votes" for one of rules that cover it, in a probabilistic way. More precisely, the probability of voting for a given rule (individual) is proportional to the fitness of that rule. Only rules covering the same examples compete with each other. Hence, this procedure effectively implements a form of niching, encouraging the evolution of several different rules, each of them covering a different part of the data space.

Generalizing/specializing crossover. The basic idea of this special kind of crossover is to generalize or specialize a given rule, depending on whether it is currently overfitting or underfitting the data [1.11].

To simplify our discussion, assume that the evolutionary algorithm follows the Michigan approach - where each individual represents a single rule - using a binary encoding. Then the generalizing / specializing crossover operators can be implemented as the logical OR and the logical AND, respectively. This is illustrated below, where the above-mentioned bitwise logical functions are used to compute the values of the bits between the two crossover points denoted by the "|" symbol.

Parents	Children generated by generalized crossover	Children generated by specialized crossover
00 | 1100 | 1	0011101	0010001
10 | 1010 | 0	1011100	1010000

Example of generalizing / specializing crossover

Generalizing/specializing-condition operator. In the previous subsection we saw how the crossover operator can be modified to generalize/ specialize a rule. However, the generalization/specialization of a rule can also be done in a way independent of crossover. Suppose, e.g., that a given individual represents a rule antecedent with two attribute-value conditions, as follows - again, there is an implicit logical AND connecting the two conditions as:

(Age > 25) (Marital_Status = "single").......... (a)

We can generalize, say, the first condition of (1) by using a kind of mutation operator that subtracts a small, randomly-generated value from 25. This might transform the rule antecedent of (a) into, say, the following one:

(Age > 21) (Marital_Status = "single").......... (b)

Rule antecedent (b) tends to cover more examples than (a), which is a kind of result that we wish in the case of a generalization operator. Another way to generalize rule antecedent (a) is simply to delete one of its conditions.

Conversely, we could specialize the first condition of rule antecedent (a) by using a kind of mutation operator that adds a small, randomly-generated value to 25. Another way to specialize (a) is, of course, to add another condition to that rule antecedent.

1.3.6 Fitness Functions for Rule Discovery

As discussed earlier, ideally the discovered rules should: (a) have a high predictive accuracy (b) be comprehensible and (c) be interesting. In this subsection we discuss how these rule quality criteria can be incorporated in a fitness function. To simplify our discussion, throughout this subsection we will again assume that the GA follows the Michigan approach - i.e. an individual represents a single rule. However, the basic ideas discussed below can be easily adapted to GAs following the Pittsburgh approach, where an individual represents a rule set.

Let a rule be of the form: IF A THEN C, where A is the antecedent (a conjunction of conditions) and C is the consequent (predicted class), as discussed earlier. A very simple way to measure the predictive accuracy of a rule is to compute the so-called confidence factor (CF) of the rule, defined as:

$$CF = |A \cup C| / |A|,$$

where $|A|$ is the number of examples satisfying all the conditions in the antecedent A and $|A \cup C|$ is the number of examples that both satisfy the antecedent A and have the class predicted by the consequent C. For instance, if a rule covers 10 examples (i.e. $|A| = 10$), out of which 8 have the class predicted by the rule (i.e. $|A\&C| = 8$) then the CF of the rule is CF = 80%.

Unfortunately, such a simple predictive accuracy measure favors rules overfitting the data. For instance, if $|A| = |A \& C| = 1$ then the CF of the rule is 100%. However, such a rule is most likely representing an idiosyncrasy of a particular training example, and probably will have a poor predictive accuracy on the test set. A solution for this problem is described next.

The predictive performance of a rule can be summarized by a 2 x 2 matrix, sometimes called a confusion matrix. To interpret this, recall that A denotes a rule antecedent and C denotes the class predicted by the rule. The class predicted for an example is C if and only if the example satisfies the rule antecedent. The labels in each quadrant of the matrix have the following meaning:

TP = True Positives = Number of examples satisfying A and C

FP = False Positives = Number of examples satisfying A but not C

FN = False Negatives = Number of examples not satisfying A but satisfying C

TN = True Negatives = Number of examples not satisfying A nor C

Clearly, the higher the values of TP and TN, and the lower the values of FP and FN, the better the rule.

1.4 Multi-Objective Optimization and Rule Mining Problems

It is always difficult to find out a single solution for a multi-objective problem. So it is natural to find out a set of solutions depending on non-dominance criterion [1.7]. At the time of taking a decision, the solution that seems to fit better depending on the circumstances can be chosen from the set of these candidate solutions. A solution, say a, is said to be dominated by another solution, say b, if and only if the solution b is better or equal with respect to all the corresponding objectives of the solution a, and b is strictly better in at least one objective. Here the solution b is called a *non-dominated* solution. So it will be helpful for the decision-maker, if we can find a set of such non-dominated solutions. Vilfredo Pareto suggested this approach of solving the multi objective problem. Optimization techniques based on this approach are termed as Pareto optimization techniques.

Based on this idea, several genetic algorithms were designed to solve multi-objective problems [1.7]. Multiple-Objective Genetic Algorithm (MOGA) [1.7] is one of them. Here the *chromosomes* are selected (using standard selection scheme, e.g. *roulette wheel* selection) using the fitness value. Fitness value is calculated using their ranks, which are calculated from the non-dominance property of the chromosomes. The ranking step tries to find the non-dominated solutions, and those solutions are ranked as one. Among the rest of the chromosomes, if p_i individuals dominate a chromosome then its rank is assigned as $1+p_i$. This process continues till all the chromosomes are ranked. Then fitness is assigned to the chromosomes such that the chromosomes having the smallest rank gets the highest fitness and the chromosomes having the same rank gets the same fitness. After assigning the fitness to the chromosomes, *selection, replacement, crossover* and *mutation* operators are applied to get a new set of chromosomes, as in standard GAs.

As mentioned earlier, in the present work we used the *comprehensibility* and the *interestingness* measure of rules in addition to predictive accuracy (which is already discussed in the previous section) as objectives of multi-objective GAs. Let us discuss them here. It is very difficult to quantify understandability or comprehensibility. A careful study of an association rule will infer that if the number of conditions involved in the antecedent part is less, the rule is more comprehensible. To reflect this behavior, an expression was derived as $comp = N$-(number of conditions in the antecedent part) [1.11]. This expression serves well for the classification rule generation where the number of attributes in the consequent part is always one. Since, in the association rules, the consequent part may contain more than one attribute, this expression is not suitable for the association rule mining. We require an expression where the number of attributes involved in both the parts of the rule has some effect. The following expression can be used to quantify the comprehensibility of an association rule,

$$Comprehensibility = log(1+|C|) \ / \ log(1+|A \cup C|).$$

Here, $|C|$ and $|A \cup C|$ are the number of attributes involved in the consequent part and the total rule, respectively.

Since association rule mining is a part of data mining process that extracts some hidden information, it should extract only those rules that have a comparatively less occurrence in the entire database. Such a surprising rule may be more interesting to the users; which again is difficult to quantify. For classification rules it can be defined by information gain theoretic measures [1.11]. This way of measuring interestingness for the association rules will become computationally inefficient. For finding interestingness the data set is to be divided based on each attribute present in the consequent part. Since a number of attributes can appear in the consequent part and they are not predefined, this approach may not be feasible for association rule mining. So a new expression is defined which uses only the support count of the antecedent and the consequent parts of the rules, and is defined as

$$Interestingness = [SUP(A \cup C)/SUP(A)] \times [SUP(A \cup C)/SUP(C)] \times \\ [1 - (SUP(A \cup C)/|D|)].$$

Here $|D|$ is the total number of records in the database.

This expression contains three parts. The first part, $[SUP(A \cup C)/SUP(A)]$, gives the probability of generating the rule depending on the antecedent part, the second part, $[SUP(A \cup C)/SUP(C)]$, gives the probability of generating the rule depending on the consequent part, and $(SUP(A \cup C)/|D|)$ gives the probability of generating the rule depending on the whole data-set. So complement of this probability will be the probability of **not generating** the rule. Thus, a rule having a very high support count will be measured as less interesting.

1.4.1 Model of Ghosh and Nath

In this work [1.12] we tried to solve the association rule-mining problem with a Pareto based genetic algorithm. The first task for this is to represent the possible rules as chromosomes, for which a suitable encoding/decoding scheme is required. For this, two approaches can be adopted. In the *Pittsburgh approach* each chromosome represents a set of rules, and this approach is more suitable for classification rule mining; as we do not have to decode the consequent part and the length of the chromosome limits the number of rules generated. The other approach is called the *Michigan approach* where each chromosome represents a separate rule. In the original Michigan approach we have to encode the antecedent and consequent parts separately; and thus this maybe an efficient way from the point of space utilization since we have to store the empty conditions as we do not known a priori which attributes will appear in which part. So we followed a new approach that is better than this approach from the point of storage requirement. With each attribute we associate two extra tag bits. If these two bits are **00** then the attribute next to these two bits appears in the antecedent part and if it is **11** then the attribute appears in the consequent part. And the other two combinations, **01** and **10** will indicate the absence of the attribute in either of these parts. So the above rule will look like **00A 11B 00C 01D 11E 00F**. In this way we can handle variable length rules with more storage efficiency, adding only an overhead of $2k$ bits, where k is the number of attributes in the database.

The next step is to find a suitable scheme for encoding/decoding the rules to/from binary chromosomes. Since the positions of attributes are fixed, we need not store the name of the attributes. We have to encode the values of different attribute in the chromosome only. For encoding a categorical valued attribute, the market basket encoding scheme is used. As discussed earlier this scheme is not suitable for numeric valued attributes. For a real valued attribute their binary representation can be used as the encoded value. The range of value of that attribute will control the number of bits used for it. Decoding will be simply the reverse of it. The length of the string will depend on the required accuracy of the value to be encoded. Decoding can be performed as:

$$Value = Minimum\ value + (maximum\ value - minimum\ value) \times$$
$$((\sum(2^{i-1} \quad \times \quad i^{th}\ \text{bit value}))/(2^n\text{-}1))$$

Where $1 \leq i \leq n$ and n is the number of bits used for encoding; and *minimum* & *maximum* are minimum and maximum values of the attribute.

Using these encoding schemes, values of different attributes can be encoded into the chromosomes. Since in the association rules an attribute may be involved with different relational operators [1.11], it is better to encode them also within the rule itself. For example, in one rule a numeric attribute A may be involved as $A \geq value_1$, but in another rule it may be involved as $A \leq value_2$. Similarly, a categorical attribute may be involved with either *equal to* ($=$) or *not equal to* ($\neq$). To handle this situation we used another bit to indicate the operators involved with the attribute. *Equality* and *not equality* are not considered with the numerical attribute. In this way the whole rule can be represented as a binary string, and this binary string will represent one chromosome or a possible rule.

After getting the chromosomes, various genetic operators can be applied on it. Presence of large number of attributes in the records will results in large chromosomes, thereby needing multi-point crossover.

There are some difficulties to use the standard multi-objective GAs for association rule mining problems. In case of rule mining problems, we need to store a set of better rules found from the database. If we follow the standard genetic operations only, then the final population may not contain some rules that are better and were generated at some intermediate generations. It is better to keep these rules. For this task, a separate population is used [1.1]. In this population no genetic operation is performed. It will simply contain only the non-dominated chromosomes of the previous generation. The user can fix the size of this population. At the end of first generation, it will contain the non-dominated chromosomes of the first generation. After the next generation, it will contain those chromosomes, which are non-dominated among the current population as well as among the non-dominated solutions till the previous generation.

The approach will work as follows:

1. Load a sample of records from the database that fits in the memory.
2. Generate N chromosomes randomly.
3. Decode them to get the values of the different attributes.
4. Scan the loaded sample to find the support of antecedent part, consequent part and the rule.
5. Find the confidence, comprehensibility and interestingness values.
6. Rank the chromosomes depending on the non-dominance property.
7. Assign fitness to the chromosomes using the ranks, as mentioned earlier.
8. Bring a copy of the chromosomes ranked as 1 into a separate population, and store them if they are non-dominated in this population also. If some of the existing chromosomes of this population become dominated, due to this insertion, then remove the dominated chromosomes from this population.
9. Select the chromosomes, for next generation, by roulette wheel selection scheme using the fitness calculated in Step 7.
10. Replace all chromosomes of the old population by the chromosomes selected in Step 9.
11. Perform multi-point crossover and mutation on these new individuals.
12. If the desired number of generations is not completed, then go to Step 3.
13. Decode the chromosomes in the final stored population, and get the generated rules.

1.5 Discussion and Research Directions

We have begun our discussion of data mining and knowledge discovery by identifying three desirable properties of discovered knowledge. These properties are predictive accuracy, comprehensibility and interestingness. We believe a promising research direction is to design evolutionary algorithms which aim at discovering truly interesting rules. Clearly, this is much easier said than done. The major problem is that rule interestingness is a complex concept, involving both objective and subjective aspects. Almost all the fitness functions currently used in evolutionary algorithms for data mining focus on the objective aspect of rule quality, and in most cases only predictive accuracy and rule comprehensibility are taken into account. However, these two factors alone do not guarantee rule interestingness, since a highly-accurate, comprehensible rule can still be uninteresting, if it corresponds to a piece of knowledge previously known by the user.

Concerning data mining tasks, which correspond to kinds of problems to be solved by data mining algorithms, in this chapter we have focused on the rule mining task only, due to space limitations. However, many of the ideas and concepts discussed here are relevant to other data mining tasks involving prediction, such as the dependence modelling task.

We have discussed several approaches to encode prediction (IF-THEN) rules into the genome of individuals, as well as several genetic operators designed specifically for data mining purposes. A typical example is the use of generalizing/specializing crossover. We believe that the development of new data mining-oriented operators is important to improve the performance of evolutionary algorithms in data mining and knowledge discovery. Using this kind of operator makes evolutionary algorithms endowed with some "knowledge" about what kind of genome-modification operation makes sense in data mining problems. The same argument holds for other ways of tailoring evolutionary algorithms for data mining, such as developing data mining-oriented individual representations.

We have also discussed the use of multi-objective evolutionary algorithms for rule mining. Use of other multi-objective EAs for rule mining needs to be investigated.

References

1.1 Adamo, J. M. (2001): Data mining for association rules and sequential patterns. Springer-Verlag, USA
1.2 Agrawal, R., Imielinski, T., Swami, A. (1993): Mining association rules between sets of items in large databases. Proc. 1993 Int. Conf. Management of Data (SIGMOD-93), 207-216
1.3 Anglano, C., Giordana, A., Lo Bello, G., Saitta, L. (1998): Coevolutionary, distributed search for inducing concept descriptions. Lecture Notes in Artificial Intelligence 1398. ECML-98: Proc. 10th Europ. Conf. Machine Learning, 422-333. Springer-Verlag
1.4 Banzhaf, W., Nordin, P., Keller, R. E., Francone, F. D. (1998): Genetic programming ~ an introduction: on the automatic evolution of computer programs and its applications. Morgan Kaufmann
1.5 Cios, K., Pedrycz, W., Swiniarski, R. (2000): Data mining methods for knowledge discovery. Kluwer Academic Publishers
1.6 Cooley, R., Mobasher, B., Srivastava, J. (1997): Web mining: information and pattern discovery on the world wide web. Proc. Of the 9th IEEE International Conference on Tools with Artificial Intelligence
1.7 Deb, K. (2001): Multi-objective optimization using evolutionary algorithms. John Wiley & Sons
1.8 Duda, R. G., Hart, P. E., Stork, D. G. (2001): "Pattern classification". John Wiles & Sons

1.9 Fayyad, U. M., Piatetsky-Shapiro, G., Smyth. P. (1996): From data mining to knowledge discovery: an overview. In: Fayyad UM, Piatetsky-Shapiro G, Smyth P and Uthurusamy R. Advances in Knowledge Discovery & Data Mining, 1-34. AAAI/MIT

1.10 Flockhart, I. W., Radcliffe, N. J. (1995): GA-MINER: parallel data mining with hierarchical genetic algorithms - final report. EPCC-AIKMS-GA-MINER-Report 1.0. University of Edinburgh, UK

1.11 Freitas, A. A. (2002): Data mining and knowledge discovery with evolutionary algorithms. Springer-Verlag

1.12 Ghosh, A., Nath, B. (2004): "Multi-objective rule mining using genetic algorithms". Information Sciences (in press).

1.13 Giordana, A. Neri, F. (1995): Search-intensive concept induction. Evolutionary Computation **3**, 375-416

1.14 Giordana, A., Saitta, L., Zini, F. (1994): Learning disjunctive concepts by means of genetic algorithms. Proc. 10th Int. Conf. Machine Learning (ML-94), 96-104. Morgan Kaufmann

1.15 Goldberg, D. E. (1989): Genetic algorithms in search, optimization and machine learning. Addison-Wesley

1.16 Han, J., Kamber, M. (2001): Data mining: concept and techniques. Morgan Kauffman Publisher

1.17 Han, J., Jamil, H., Lu, Y., Chan, L., Liao, Y., Pei, J.: DNA-Miner: a system prototype for mining DNA sequences, Electronic Edition, ACM, School of Computing Science, Simon Fraser University, Canada; Dept. of Computer Science, Mississipi State University, USA

1.18 Jain, A., Zongker, D. (1997): Feature selection: evaluation, application, and small sample performance. IEEE Transactions on Pattern Analysis and Machine Intelligence (PAMI), 19, 153-158

1.19 Janikow C. Z. (1993): A knowledge-intensive genetic algorithm for supervised learning. Machine Learning **13**, 189-228

1.20 Koza, J. R. (1992): Genetic programming: on the programming of computers by means of natural selection. MIT Press

1.21 Mahfoud, S. W. (1995): Niching methods for genetic algorithms. Ph.D. Thesis. Univ. of Illinois at Urbana-Champaign. IlliGAL Report No. 95001

1.22 Michalewicz, Z. (1996): Genetic algorithms + data structures = evolution programs. 3rd Ed. Springer-Verlag

1.23 Mobasher, B., Jain, N., Han, E., Srivastava, J. (1996): Web mining: pattern discovery from world wide web transactions, Technical Report TR-96050, Dept. of Computer Science, University of Minnesota, Minneapolis

1.24 Pyle, D. (1999): Data preparation for data mining. Morgan Kaufmann

1.25 Ramakrishnan, N., Grama, A. Y. (2001): Mining scientific data, advances in computers, **55**, 119—169

1.26 Su, S., Cook, D. J., Holder, L. B. (1999): Application of knowledge discovery in molecular biology: identifying structural regularities in proteins. Proc. Of Pacific Symposium of Biocomputing, 190—201.

1.27 Weiss, S. M., Indurkhya, N. (1998): Predictive data mining: a practical guide. Morgan Kaufmann

2. Strategies for Scaling Up Evolutionary Instance Reduction Algorithms for Data Mining

Jose Ramon Cano[1], Francisco Herrera[2], and Manuel Lozano[2]

[1] Dept. of Electronic Engineering, Computer Systems and Automatics, Escuela Superior de La Rabida, University of Huelva, 21819, Huelva, Spain
jose.cano@diesia.uhu.es
[2] Dept. of Computer Science and Artificial Intelligence, University of Granada, 18071, Granada, Spain
herrera@decsai.ugr.es, lozano@decsai.ugr.es

Abstract: Evolutionary algorithms are adaptive methods based on natural evolution that may be used for search and optimization. As instance selection can be viewed as a search problem, it could be solved using evolutionary algorithms.

In this chapter, we have carried out an empirical study of the performance of CHC as representative evolutionary algorithm model. This study includes a comparison between this algorithm and other non-evolutionary instance selection algorithms applied in different size data sets to evaluate the scaling up problem. The results show that the stratified evolutionary instance selection algorithms consistently outperform the non-evolutionary ones. The main advantages are: better instance reduction rates, higher classification accuracy and reduction in resources consumption.

2.1 Introduction

Advances in digital and computer technology that have led to the huge expansion of the Internet means that massive amounts of information and collection of data have to be processed. Due to the enormous amounts of data, much of the current research is based on scaling up [2.5] Data Mining (DM) ([2.1, 2.20, 2.22]) algorithms. Other research has also tackled scaling down data. The main problem of scaling down data is how to select the relevant data. This task is carried out in the data preprocessing phase in a Knowledge Discovery in Databases (KDD) process.

Our attention is focused on Data Reduction (DR) [2.16] as preprocessing task, which can be achieved in many ways: by selecting features [2.15], by making the feature-values discrete [2.8], and by selecting instances([2.13]). We led our study to Instance Selection (IS) as DR mechanism ([2.3, 2.17, 2.18]), where we reduce the number of rows in a data set (each row represents and instance). IS can follow different strategies (see Fig. 2.1): sampling, boosting, prototype selection (PS), and active learning. We are going to study the IS from the PS perspective.

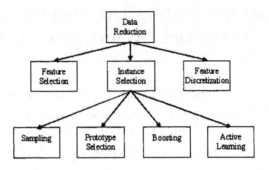

Fig. 2.1. Data reduction strategies

IS mechanisms have been proposed to choose the most suitable points in the data set to become instances for the training data set used by a learning algorithm. IS has been studied previously in the literature using different approaches, in particular by means of Genetic Algorithms (GA) ([2.11]) as PS approach. For example, in [2.14], a GA is used for carry out a k-nearest neighbor edition.

Evolutionary Algorithms (EAs) ([2.2]) are general-purpose search algorithms that use principles inspired by natural genetic populations to evolve solutions to problems. The basic idea is to maintain a population of chromosomes, which represent plausible solutions to the problem, which evolves over time through a process of competition and controlled variation. EAs in general and GAs in particular have been used to solve the IS problem, with promising results ([2.14]).

The issue of scalability and the effect of increasing the size of data sets are always present in the algorithm behavior. This scaling up drawback appears in EAs due to the increasing of the chromosome's size, which reduces the EAs convergence capabilities.

To avoid this drawback we offer a combination of EAs and a stratified strategy. In large size we can't evaluate the algorithms over the complete data set so the stratification is a way to carry out the executions. Combining the subsets selected from the strata we can obtain the subset selected for the whole initial data set. The stratification reduces the data set size, while EAs select the most representative prototype per stratus.

The aim of this chapter is to study the application of a representative and efficient EA model for data reduction, the CHC algorithm in IS ([2.6, 2.4]), and to compare it with non-evolutionary instance selection algorithms (hereafter referred to as classical ones) following a stratified strategy.

To address this, we have carried out a number of experiments increasing the complexity and the data set size.

In order to do that, this chapter is set up as follows. In Section 2.2, we introduce the main ideas about IS, describing the processes which IS algorithms take part, and we also summarize the classical IS algorithms used in this study. In Section 2.3, we introduce the foundations of EAs and summarize the main features of them, giving details of how EAs can be applied to the IS problem in large size data sets. Section 2.4 is dedicated to the Scaling Up problem and we present our proposal solution. In Section 2.5, we explain the methodology used in the experiments. Section 2.6 deals with the results and the analysis of medium and large size data sets. Finally, in Section 2.7, we point out our conclusion.

2.2 Instance Selection on Data Reduction

In this section we describe the strategy which IS takes part in, as a DR mechanism, as well as a summary of classical IS algorithms.

2.2.1 Instance Selection

In IS we want to isolate the smallest set of instances which enable us to predict the class of a query instance with the same (or higher) accuracy as the original set [2.16]. By reducing the "useful" data set size, which can reduce both space and time complexities of subsequent processing phases. One can also hope to reduce the size of formulas obtained by a subsequent induction algorithm on the reduced and less noise data sets. This may facilitate interpretation tasks.

IS raises the problem of defining relevance for a prototype subset. From the statistical viewpoint, relevance can be partly understood as the contribution to the overall accuracy, that would be e.g. obtained by a subsequent induction. We emphasize that removing instances does not necessarily lead to a degradation of the results: we have observed experimentally that a little number of instances can have performances comparable to those of the whole sample, and sometimes higher. Two reasons come to mind to explain such an observation:

- First, some noises or repetitions in data could be deleted by removing instances.
- Second, each instance can be viewed as a supplementary degree of freedom. If we reduce the number of instances, we can sometimes avoid over-fitting situations.

2.2.2 Instance Selection for Prototype Selection

There may be situations in which there is too much data and this data in most cases is not equally useful in the training phase of a learning algorithm.

Instance selection mechanisms have been proposed to choose the most suitable points in the data set to become instances for the training data set used by a learning algorithm.

Fig. 2.2 shows a general framework for the application of an IS algorithm for PS. Starting from the data set, D, the PS algorithm finds a suitable set, Prototype Subset Selected (PSS), then a learning or DM algorithm is applied to evaluate each subset selected (1-nearest neighbor in our case) to test the quality of the subset selected. This model is assessed using the test data set, TS.

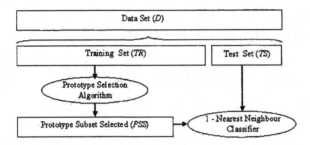

Fig. 2.2. Prototype selection strategy

2.2.3 Overview of Instance Selection Algorithms

Historically, IS has been mainly aimed at improving the efficiency of the Nearest Neighbor (NN) classifier. The NN algorithm is one of the most venerable algorithms in machine learning. This algorithm calculates the Euclidean distance (possibly weighted) between an instance to be classified and each training-neighboring instance. The new instance obtained is assigned to the class of the nearest neighboring one. More generally, the k-nearest neighbors (k-NN) are computed, and the new instance is assigned to the most frequent class among these k neighbors. The k-NN classifier was also widely used and encouraged by early theoretical results related to its Bayes error generalization.

However, from a practical point of view, the k-NN algorithm is not suitable for dealing with very large sets of data due to the storage requirements it demands and the computational costs involved. In fact, this approach requires the storage of all the instances in memory. Early research in instance selection firstly tried to reduce storage size. Taking as reference our study in [2.4] we select the most effective classic algorithms to evaluate them.

The algorithms used in this study will be:

Methods based on nearest neighbor rules.

- Cnn [2.12] - It tries to find a consistent subset, which correctly classifies all of the remaining points in the sample set. However, this algorithm will not find a minimal consistent subset.
- Ib2 [2.13] - It is similar to Cnn but using a different selection strategy.
- Ib3 [2.13] - It outperforms Ib2 introducing the acceptable instance concept to carry out the selection.

Methods based on ordered removal.

- Drop1 [2.21] - Essentially, this rule tests to see if removing an instance would degrade leave-one-out cross-validation generalization accuracy, which is an estimate of the true generalization ability of the resulting classifier.
- Drop2 [2.21] - Drop2 changes the order of removal of instances. It initially sorts the instances in TR by the distance to their nearest enemy (nearest instance belonging to another class). Instances are then checked for removal beginning at the instance furthest from its nearest enemy. This tends to remove instances furthest from the decision boundary first, which in turn increases the chance of retaining border points.
- Drop3 [2.21] - Drop3 uses a noise filtering pass before sorting the instances in TR. This is done using the rule: Any instance not classified by its k-nearest neighbors is removed.

2.3 Evolutionary Instance Selection Algorithms

EAs ([2.2]) are stochastic search methods that mimic the metaphor of natural biological evolution. All EAs rely on the concept of a population of individuals (representing search points in the space of potential solutions to a given problem), which undergo probabilistic operators such as mutation, selection, and (sometimes) recombination to evolve towards increasingly better fitness values of the individuals.

Most of the success of EAs is due to their ability to exploit the information accumulated about an initially unknown search space. This is their key feature, particularly in large, complex, and poorly understood search spaces, where classical search tools (enumerative, heuristic, etc.) are inappropriate. In such cases they offer a valid approach to problems requiring efficient and effective search techniques. Recently EAs have been widely applied to KDD and DM ([2.9, 2.10]).

In this section we firstly present the key-points of their application to our problem as well as the representation and the fitness function, and then describe the EA (CHC [2.6]) used in this study, according to the best results obtained by CHC in the study presented in [2.4].

In the Section 2.3.2 we describe the model of EA that will be used in this chapter as evolutionary instance selection algorithm. CHC is a classical model that introduces different features to obtain a trade-off between exploration and exploitation.

2.3.1 Evolutionary Instance Selection: Key Points

EAs may be applied to the IS problem, because it can be considered as a search problem.

The application of EAs to IS is accomplished by tackling two important issues: the specification of the representation of the solutions and the definition of the fitness function.

Representation. Let's assume a data set denoted TR with n instances. The search space associated with the instance selection is constituted by all the subsets of TR. Then, the chromosomes should represent subsets of TR. This is accomplished by using a binary representation. A chromosome consists of n genes (one for each instance in TR) with two possible states: 0 and 1. If the gene is 1, then its associated instance is included in the subset of TR represented by the chromosome. If it is 0, then this does not occur.

Fitness function. Let PSS be a subset of instances of TR to evaluate and be coded by a chromosome. We define a fitness function that combines two values: the classification performance (clas_per) associated with PSS and the percentage of reduction (perc_red) of instances of PSS with regards to TR:

$$Fitness(\text{PSS}) = \alpha \cdot \text{clas\_rat} + (1 - \alpha) \cdot \text{perc\_red}. \qquad (2.1)$$

The 1-NN classifier (Section 2.2.3) is used for measuring the classification rate, clas_rat, associated with PSS. It denotes the percentage of correctly classified objects from TR using only PSS to find the nearest neighbor. For each object y in TR, the nearest neighbor is searched for amongst those in the set PSS $\setminus \{y\}$. Whereas, perc_red is defined as:

$$\text{perc\_red} = 100 \cdot (|\text{TR}| - |\text{PSS}|)/|\text{TR}|. \qquad (2.2)$$

The objective of the EAs is to maximize the fitness function defined, i.e., maximize the classification performance and minimize the number of instances obtained. In the experiments presented in this chapter, we have considered the value $\alpha = 0.5$ in the fitness function, as per a previous experiment in which we found the best trade-off between precision and reduction with this value.

2.3.2 The CHC Algorithm

During each generation the CHC develops the following steps:

1. It uses a parent population of size n to generate an intermediate population of n individuals, which are randomly paired and used to generate n potential offspring.
2. Then, a survival competition is held where the best n chromosomes from the parent and offspring populations are selected to form the next generation.

CHC also implements a form of heterogeneous recombination using HUX, a special recombination operator. HUX exchanges half of the bits that differ between parents, where the bit position to be exchanged is randomly determined. CHC also employs a method of incest prevention. Before applying HUX to two parents, the Hamming distance between them is measured. Only those parents who differ from each other by some number of bits (mating threshold) are mated. The initial threshold is set at L/4, where L is the length of the chromosomes. If no offspring are inserted into the new population then the threshold is reduced by 1.

No mutation is applied during the recombination phase. Instead, when the population converges or the search stops making progress (i.e., the difference threshold has dropped to zero and no new offspring are being generated which are better than any members of the parent population) the population is re-initialized to introduce new diversity to the search. The chromosome representing the best solution found over the course of the search is used as a template to re-seed the population. Re-seeding of the population is accomplished by randomly changing 35% of the bits in the template chromosome to form each of the other n-1 new chromosomes in the population. The search is then resumed.

2.4 The Scaling up Problem. The Stratified Approach

In this section we point our attention in the Scaling Up problem and finally we describe our proposal, the combination of the stratified strategy with the evolutionary instance selection.

2.4.1 Scaling up and Stratification

The algorithms we have studied, both classical and evolutionary, are affected when the size of the data set increases. The main difficulties they have to face are as follows:

– Efficiency. The efficiency of IS algorithms is at least $O(n^2)$, where n is the size of the data set. Most of them present an efficiency order greater than $O(n^2)$. When the size increases, the time needed by each algorithm also increases.

- Resources. Most of the algorithms assessed need to have the complete data set stored in memory to carry out their execution. If the size of the problem is too big, the computer would need to use the disk as swap memory. This loss of resources has an adverse effect on efficiency due to the increased access to the disk.
- Representation. EAs are also affected by representation, due to the size of their chromosomes. When the size of these chromosomes is too big, then it increases the algorithm convergence difficulties.

To avoid these drawback we led our experiments towards a stratified strategy. This strategy divides the initial data set in strata. The strata are disjoints sets with equal class distribution. We evaluate the algorithm over each stratus to carry out the data selection and finally we reunite the partial subsets to conform the final one.

In the following section (Fig. 2.3) we describe the use of the stratified strategy combined with EA.

2.4.2 Evolutive Algorithms and Stratification Strategy

Following the stratified strategy, initial data set D is divided into t disjoint sets D_j, strata of equal size, $D_1, D_2, ...,$ and D_t. We maintain class distribution within each set in the partitioning process.

Prototype selection algorithms (classical or evolutionary) are applied to each D_j obtaining a subset selected DS_j, as we can see in Fig. 2.3.

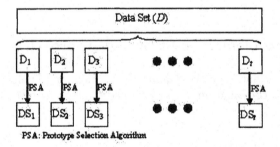

Fig. 2.3. Combination of prototype selection algorithms and stratified strategy

In Fig. 2.2, the PSS is obtained by the PS algorithm, applied on TR. In the stratified strategy, the PS algorithm is applied in each D_j to obtain its DS_j associated. PSS in stratified strategy is obtained using the DS_j (see Eq. (2.3)) and it is called Stratified Prototype Subset Selected (SPSS).

$$SPSS = \bigcup_{j \in J} DS_j, J \subset \{1, 2, ..., t\} \qquad (2.3)$$

The test set TS will be the TR complementary one in D.

$$TR = \bigcup_{j \in J} D_j, J \subset \{1, 2, ..., t\} \tag{2.4}$$

$$TS = D \setminus TR \tag{2.5}$$

Our specific model will be described in Section 2.5.2.

2.5 Experimental Methodology

We have carried out our study of IS problem using two size problems: medium and large. We intend to study the behavior of the algorithms when the size of the problem increases. Section 2.5.1 describes the data sets used and introduces the parameters associated with the algorithms, Section 2.5.2 introduces the stratification and partition of the data sets that were considered for applying the algorithms, and finally, in Section 2.5.3 we describe the table contents that show the results.

2.5.1 Data Sets and Parameters

The data sets used are shown in Table 2.1 and 2.2. They can be found in the UCI Repository (http://kdd.ics.uci.edu/).

Table 2.1. Medium size data sets

Data Set	Num. Instances	Num. Features	Num. Classes
Pen-Based Recognition	10992	16	10
Satimage	6435	36	6
Thyroid	7200	21	3

Table 2.2. Large size data set

Data Set	Num. Instances	Num. Features	Num. Classes
Adult	30132	14	2

The parameters used are shown in Table 2.3.

Table 2.3. Parameters

Algorithm	Parameters
Ib3	Acept. Level=0.9, Drop Level=0.7
CHC	Population=50, Evaluations=10000

2.5.2 Partitions and Stratification: A Specific Model

We have evaluated each algorithm in a ten fold cross validation process. In the validation process TR_i, i=1, ..., 10 is a 90% of D and TS_i its complementary 10% of D.

In our experiments we have evaluated the PS algorithms following two perspectives for the ten fold cross validation process.

In the first one, we have executed the PS algorithms as we can see in Fig. 2.4. We call it Ten fold cross validation classic (Tfcv classic). The idea is use this result as baseline versus the stratification ones.

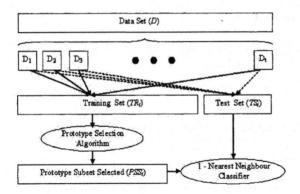

Fig. 2.4. Prototype selection strategy in Tfcv classic

In Tfcv classic the subsets TR_i and TS_i, i=1, ..., 10 are obtained as the Eqs. (2.6) and (2.7) indicate:

$$TR_i = \bigcup_{j \in J} D_j, J = \{j/1 \le j \le b \cdot (i-1) \ and \ (i \cdot b) + 1 \le j \le t\} \quad (2.6)$$

$$TS_i = D \setminus TR_i \quad (2.7)$$

where t is the number of strata, and b is the number of strata grouped ($b = t/10$).

Each PSS_i is obtained by the PS algorithm applied to TR_i subset.

The second way is to execute the PS algorithms in a stratified process as the Fig. 2.5 shows. We call it Ten fold cross validation strat (Tfcv strat).

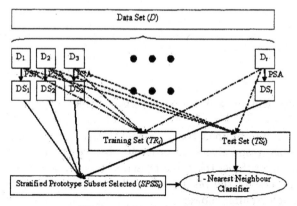

PSA: Prototype Selection Algorithm

Fig. 2.5. Prototype selection strategy in Tfcv strat

In Tfcv strat each TR_i is defined as we can see in Eq. (2.6), by means of the union of D_j subsets (see Fig. 2.5).

In Tfcv strat (see Fig. 2.5) $SPSS_i$ is generated using the DS_j (see Eq. (2.8)).

$$SPSS_i = \bigcup_{j \in J} DS_j, J = \{j/1 \leq j \leq b \cdot (i-1) \ and \ (i \cdot b) + 1 \leq j \leq t\}(2.8)$$

$SPSS_i$ contains the instances selected by PS algorithms in TR_i following the stratified strategy.

The subset TS_i is defined by means the Eq. (2.7). Both, TR_i and TS_i are generated in the same way in Tfcv classic and Tfcv strat.

For each data set we have employed the following partitions and number of strata:

Table 2.4. Stratification in medium size data sets

Pen-Based Recognition	Satimage	Thyroid
10 Strata	10 Strata	10 Strata
30 Strata	30 Strata	30 Strata

Table 2.5. Stratification in large size data set

Adult
10 Strata
50 Strata
100 Strata

2.5.3 Table of Results

In the following section we will present the structure of tables where we present the results.

Our table shows the results obtained by the classical and evolutionary instance selection algorithms, respectively. In order to observe the level of robustness achieved by all algorithms, the table presents the average in the ten fold cross validation process of the results offered by each algorithm in the data sets evaluated. Each column shows:

- The first column shows the name of the algorithm. In this column the name is followed by the sort of validation process (Tfcv strat and the number of strata, or Tfcv classic meaning ten fold cross-validation process classic).
- The second column contains the average execution time associated to each algorithm. The algorithms have been run in a Pentium 4, 2.4 Ghz, 256 RAM, 40 Gb HD.
- The third column shows the average reduction percentage from the initial training sets.
- The fourth column contains the training accuracy associated to the prototype subset selected.
- The fifth column contains the test accuracy of the PS algorithms selection.

2.6 Experimental Study

In this section we present the results obtained in the evaluation of medium and large data sets and their analysis.

2.6.1 Medium Size Data Sets

The following conclusions about the IS algorithms for PS can be made studying Table 2.6:

- In Table 2.6 we can see that the stratification strategy reduces significantly the execution time.
- Stratified strategy affects in different manner to the accuracy rates associated to the classic algorithms. Some of them, like Ib2, Ib3 or Cnn, increase their accuracy, but other group (Drop1, Drop2 and Drop) reduces it.

Table 2.6. Prototype selection for pen-based recognition data set

	Exec. Time(sec)	% Reduction	1-NN %Ac.Trn	1-NN %Ac.Test
1-NN	66		99.36%	99.39%
Cnn Tfcv classic	4	98.04%	84.85%	85.69%
Cnn Tfcv strat 10	0.20	91.81%	93.78%	95.43%
Cnn Tfcv strat 30	0.07	82.48%	97.51%	98.63%
Drop1 Tfcv classic	374	98.45%	86.23%	86.02%
Drop1 Tfcv strat 10	2	99.86%	57.14%	22.00%
Drop1 Tfcv strat 30	0.23	99.70%	68.96%	38.90%
Drop2 Tfcv classic	318	97.69%	91.03%	91.06%
Drop2 Tfcv strat 10	1.9	98.50%	52.98%	62.92%
Drop2 Tfcv strat 30	0.27	95.37%	81.83%	78.08%
Drop3 Tfcv classic	391	98.07%	90.33%	90.05%
Drop3 Tfcv strat 10	2.1	99.66%	53.12%	40.91%
Drop3 Tfcv strat 30	0.23	98.60%	90.51%	57.53%
Ib2 Tfcv classic	2	98.49%	74.20%	75.04%
Ib2 Tfcv strat 10	0.1	94.31%	93.73%	91.41%
Ib2 Tfcv strat 30	0.03	88.34%	96.25%	97.80%
Ib3 Tfcv classic	9	96.42%	96.73%	98.00%
Ib3 Tfcv strat 10	0.2	88.34%	92.95%	98.44%
Ib3 Tfcv strat 30	0.1	83.05%	97.07%	98.63%
CHC Tfcv classic	18845	98.99%	96.29%	98.94%
CHC Tfcv strat 10	127	96.65%	98.85%	97.35%
CHC Tfcv strat 30	31	93.78%	99.69%	97.53%

- CHC and its stratified version have not been improved in their test accuracy by classic methods which offer small reduction rates. They offer the best balance between reduction and accuracy rates.
- The Stratified CHC is the one which presents the best behavior among time and resources consumption, and reduction and accuracy rates. The classic algorithm which can face to Stratified CHC is Ib3 following a Tfcv classic, which can be hard to use when the size of the problem is huge due to its resources necessities.

Table 2.7. Prototype selection for Satimage data set

	Exec. Time(sec)	% Reduction	1-NN %Ac.Trn	1-NN %Ac.Test
1-NN	36		90.33%	90.41%
Cnn Tfcv classic	5	95.93%	60.63%	61.96%
Cnn Tfcv strat 10	0.1	88.42%	68.91%	75.62%
Cnn Tfcv strat 30	0.10	79.49%	76.37%	80.46%
Drop1 Tfcv classic	206	93.66%	84.29%	81.68%
Drop1 Tfcv strat 10	1.3	98.03%	83.18%	38.12%
Drop1 Tfcv strat 30	0.13	97.89%	86.20%	30.69%
Drop2 Tfcv classic	183	83.49%	83.45%	83.51%
Drop2 Tfcv strat 10	1.2	83.55%	58.21%	79.53%
Drop2 Tfcv strat 30	0.20	80.85%	65.07%	79.06%
Drop3 Tfcv classic	301	93.25%	87.93%	81.03%
Drop3 Tfcv strat 10	1.00	96.81%	66.46%	73.02%
Drop3 Tfcv strat 30	0.13	96.65%	71.14%	57.65%
Ib2 Tfcv classic	3	96.75%	59.00%	59.59%
Ib2 Tfcv strat 10	0.20	91.87%	72.15%	66.87%
Ib2 Tfcv strat 30	0.07	85.77%	75.56%	75.81%
Ib3 Tfcv classic	22	84.66%	84.51%	86.45%
Ib3 Tfcv strat 10	0.30	78.11%	68.95%	87.50%
Ib3 Tfcv strat 30	0.10	73.71%	77.40%	87.90%
CHC Tfcv classic	2479	99.06%	89.45%	89.67%
CHC Tfcv strat 10	57	97.52%	95.23%	88.28%
CHC Tfcv strat 30	30	94.32%	97.19%	89.76%

The following conclusions can be made studying Table 2.7:

- Execution time is decreased by the stratified strategy in the same way that we saw it in Table 2.6. We can see that the stratification strategy reduces significantly the execution time.
- Stratified strategy affects in different manner to the accuracy rates associated to the classic algorithms. We can see the group conformed by the Drop family algorithms and other group with the rest of classic algorithms. The first group reduces its accuracy associated when they are evaluated following a stratification strategy while the second group increase it.

- In Satimage, CHC and its stratified version offer the best balance between reduction and accuracy rates. They have not been improved in their test accuracy by classic methods which offer small reduction rates.
- Like in Pen-Based Recognition data set, the Stratified CHC presents the best behavior among time and resources consumption, and reduction and accuracy rates.

Table 2.8. Prototype selection for thyroid data set

	Exec. Time(sec)	% Reduction	1-NN %Ac.Trn	1-NN %Ac.Test
1-NN	28		92.87%	92.74%
Cnn Tfcv classic	3	98.00%	92.50%	92.86%
Cnn Tfcv strat 10	0.10	90.72%	73.13%	90.66%
Cnn Tfcv strat 30	0.02	84.32%	76.47%	89.58%
Drop1 Tfcv classic	182	98.06%	63.47%	62.86%
Drop1 Tfcv strat 10	1.00	99.21%	80.39%	90.25%
Drop1 Tfcv strat 30	0.13	99.36%	82.22%	92.5%
Drop2 Tfcv classic	143	87.54%	91.37%	91.15%
Drop2 Tfcv strat 10	0.70	87.67%	53.40%	81.19%
Drop2 Tfcv strat 30	0.13	86.25%	61.94%	81.25%
Drop3 Tfcv classic	322	97.44%	88.82%	85.24%
Drop3 Tfcv strat 10	0.80	99.45%	80.55%	84.81%
Drop3 Tfcv strat 30	0.10	99.71%	91.17%	91.66%
Ib2 Tfcv classic	2	98.11%	92.53%	92.89%
Ib2 Tfcv strat 10	0.10	92.92%	76.50%	90.80%
Ib2 Tfcv strat 30	0.01	85.41%	76.58%	89.58%
Ib3 Tfcv classic	94	33.93%	93.22%	93.38%
Ib3 Tfcv strat 10	0.50	38.62%	93.11%	92.33%
Ib3 Tfcv strat 30	0.03	33.17%	93.70%	94.16%
CHC Tfcv classic	2891	99.83%	94.20%	91.98%
CHC Tfcv strat 10	54	99.44%	88.25%	94.01%
CHC Tfcv strat 30	33	99.16%	96.49%	93.33%

The following conclusions can be made studying Table 2.8:

- Execution time is reduced by the stratified strategy as in the Tables 2.6 and 2.7.
- In Thyroid data set, CHC and its stratified version have not been improved in their test accuracy by classic methods which offer small reduction rates. They offer the best balance between reduction and accuracy rates.
- The Stratified CHC is the one which present the best behavior among time and resources consumption, and reduction and accuracy rates.

The main conclusion that can be drawn when using medium size data sets is that Stratified CHC is the best algorithm for data reduction having both high reduction rates and accuracy, and decreasing execution time and resources consumption.

2.6.2 Large Size Data Set

Table 2.9. Prototype selection for adult data set

	Exec. Time(sec)	% Reduction	1-NN %Ac.Trn	1-NN %Ac.Test
1-NN	24		79.34%	79.24%
Cnn Tfcv classic	4	99.21%	26.40%	26.56%
Cnn Tfcv strat 10	1	97.34%	35.37%	32.02%
Cnn Tfcv strat 50	0.20	93.69%	66.51%	57.42%
Cnn Tfcv strat 100	0.02	90.09%	64.42%	58.27%
Drop1 Tfcv strat 10	44	95.09%	100.00%	25.64%
Drop1 Tfcv strat 50	1.2	94.59%	100.00%	24.96%
Drop1 Tfcv strat 100	0.15	94.49%	100.00%	24.83%
Drop2 Tfcv strat 10	48	70.33%	27.71%	61.30%
Drop2 Tfcv strat 50	0.7	68.03%	56.90%	70.27%
Drop2 Tfcv strat 100	0.13	66.96%	59.31%	71.85%
Drop3 Tfcv strat 10	41	95.57%	48.98%	63.46%
Drop3 Tfcv strat 50	0.8	95.34%	64.83%	71.19%
Drop3 Tfcv strat 100	0.11	93.71%	65.82%	70.19%
Ib2 Tfcv classic	2	99.94%	25.20%	25.14%
Ib2 Tfcv strat 10	1	99.57%	52.33%	26.89%
Ib2 Tfcv strat 50	0.1	98.66%	74.72%	45.68%
Ib2 Tfcv strat 100	0.03	94.33%	67.66%	54.30%
Ib3 Tfcv classic	210	79.42%	72.61%	74.09%
Ib3 Tfcv strat 10	3	76.69%	33.98%	70.96%
Ib3 Tfcv strat 50	0.4	73.48%	63.93%	74.36%
Ib3 Tfcv strat 100	0.05	71.21%	68.12%	71.52%
CHC Tfcv strat 10	20172	99.38%	97.02%	81.92%
CHC Tfcv strat 50	48	98.34%	93.66%	80.17%
CHC Tfcv strat 100	14	97.03%	94.28%	77.81%

We point out the following conclusions:

- If we pay attention to Table 2.9 we can see that only Cnn, Ib2 and Ib3 have been evaluated in a Tfcv classic validation. This is due to the size of the data set makes too hard to evaluate the rest of algorithms. This is one of the reason to advice the use of a stratified strategy like the one proposed by us.
- There is an important reduction in execution time due to the stratified strategy. In Stratified CHC we have reduced its execution time associated from 40.391 seconds using 3 strata to 14 seconds using 100 strata.

- Stratified CHC offers the best behavior. It presents the best reduction rates and accuracy rates, combined with a lower execution time. The fifth column in Table 2.9 shows that stratified CHC is the best algorithm offering the highest accuracy and reduction rates.
- The classical algorithms which present higher accuracy rate, offer smaller reduction rates. Those which present higher reduction rates, show minimal accuracy rates.

Clearly, when we manage large size data sets, the stratified CHC algorithm improves the behavior of the classic ones, giving the best results for scaling down data.

If we take note of Table 2.9, the initial data set which needs 24 seconds to be evaluated using 1-NN, is reduced (in 14 seconds) in 97.03% by the stratified CHC, losing less than 1.5% in accuracy rate. This situation shows that our proposal is an effective data reduction alternative to be applied in large size data sets.

Taking Table 2.9 as reference, and more concretely the Stratified CHC as the best algorithm evaluated, we can study the effect of the number of strata over the algorithm's behavior.

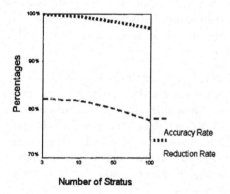

Fig. 2.6. Stratus effect on accuracy and reduction rates in Adult data set

As we can see in Fig. 2.6, when the number of strata increases, both the accuracy and reduction rate decrease. This situation is due to the selection carried out in each stratus. When the number of strata increases, the number of equivalent selected instances in different subsets also increases. This situation produces that the size of the final subset selected is bigger.

2.7 Concluding Remarks

This chapter addressed the analysis of the evolutionary instance selection by means of CHC and their use in data reduction for large data sets in KDD. We have studied the effect of the stratified strategy in the scaling up of the algorithms.

The main conclusions reached are the following:

- The Stratified Strategy reduces significantly the execution time and the resources consumed by classic and CHC algorithm. This situation offers two principal advantages: First, the evaluation of large data sets which needs too much resources is feasible, and second, the reduction in time associated to its execution.
- Stratified CHC outperform the classical algorithms, simultaneously offering two main advantages: better data reduction percentages and higher classification accuracy.
- In medium and large size data sets, classical algorithms do not present balanced behavior. If the algorithm reduces the size then its accuracy rate is poor. When accuracy increases there is no reduction.
- The increase in the number of strata can produce a small degradation in the algorithm behavior as we indicated in Fig. 2.6. The adequate number of them has to be chosen to produce a balance between time and resources consumption in one side, and reduction and accuracy rates by the other side.

Therefore, as a final concluding remark, we consider the stratified strategy combined with CHC to be a good mechanism for data reduction, facing to the problem of Scaling Up. It has become a powerful tool to obtain small selected training sets and therefore scaling down data. CHC can select the most representative instances, satisfying both the objectives of high accuracy and reduction rates. Stratified strategy permits a reduction of the search space so we can carry out the evaluation of the algorithms with acceptable execution time, and decreasing the resources necessities.

Finally, we point out that future research could be directed towards the study of hybrid strategies between classical and evolutionary instance selection algorithms.

References

2.1 Adriaans, P., Zantinge, D. (1996): Data mining. Addison-Wesley
2.2 Back, T., Fogel, D., Michalewicz, Z. (1997): Handbook of evolutionary computation. Oxford University Press
2.3 Brighton, H., Mellish, C. (2002): Advances in instance selection for instance-based learning algorithms. Data Mining and Knowledge Discovery **6**, 153–172

2.4 Cano, J.R., Herrera, F., Lozano, M. (2003): Using evolutionary algorithms as instance selection for data reduction in KDD: An experimental study. IEEE Transaction on Evolutionary Computation (In press)

2.5 Domingo, C., Gavalda, R., Watanabe, O. (2002): Adaptative sampling methods for scaling up knowledge discovery algorithms. Data Mining and Knowledge Discovery 6, 131–152

2.6 Eshelman, L. J. (1991): The CHC adaptive search algorithm: how to have safe search when engaging in nontraditional genetic recombination. (Foundations of Genetic Algorithms-1) , Rawlins, G.J.E. (Eds.), Morgan Kauffman, 265–283

2.7 Esposito, F., Malerba, D., Semeraro, G. (1997): A comparative analysis of methods for pruning decision trees. IEEE Transactions on Pattern Analysis and Machine Intelligence, 19, 476–491

2.8 Frank, E., Witten, I. H. (1999): Making better use of global discretization. (Proc. Sixteenth International Conference on Machine Learning), Bratko, I., Dzeroski, S. (Eds.), Morgan Kaufmann, 115–123

2.9 Freitas, A. A. (2002): Data mining and knowledge discovery with evolutionary algorithms. Springer-Verlag

2.10 Freitas, A.A. (2002): A survey of evolutionary algorithms for data mining and knowledge discovery. (Advances in evolutionary computation), Ghosh, A., Tsutsui, S. (Eds.), Springer-Verlag, 819–845

2.11 Goldberg, D. E. (1989): Genetic algorithms in search, optimization, and machine learning. Addison-Wesley

2.12 Hart, P. E. (1968): The condensed nearest neighbour rule. IEEE Transaction on Information Theory, 18, 431–433

2.13 Kibbler, D., Aha, D. W. (1987): Learning representative exemplars of concepts: An initial case of study. Proc. of the (Fourth International Workshop on Machine Learning) , Morgan Kaufmann, 24–30

2.14 Kuncheva, L. (1995): Editing for the k-nearest neighbors rule by a genetic algorithm. Pattern Recognition Letters, 16, 809–814

2.15 Liu, H., Motoda, H. (1998): Feature selection for knowledge discovery and data mining. Kluwer Academic Publishers

2.16 Liu, H., Motoda, H. (2001): Data reduction via instance selection. (Instance Selection and Construction for Data Mining), Liu, H., Motoda, H. (Eds.) , Kluwer Academic Publishers, 3–20

2.17 Liu, H., Motoda, H. (2002): On issues of instance selection. Data Mining and Knowledge Discovery, 6, 115–130

2.18 Reinartz, T. (2002): A unifying view on instance selection. Data mining and Knowledge Discovery, 6, 191–210

2.19 Safavian, S. R., Landgrebe, D. (1991): A survey of decision tree classifier methodology. IEEE Transaction on Systems, Man. and Cybernetics, 21, 660–674

2.20 Shanahan, J. G. (2000): Soft computing for knowledge discovery. Kluwer Academic Publishers

2.21 Wilson, D. R., Martinez, T. R. (1997): Instance pruning techniques. (Proceedings of the International Conference), Morgan Kaufmann, 403–411

2.22 Witten, I. H., Frank, E. (2000): Data mining: practical machine learning tools and techniques with Java implementations. Morgan Kaufmann

3. GAP: Constructing and Selecting Features with Evolutionary Computing

Matthew G. Smith and Larry Bull

Faculty of Computing, Engineering and Mathematical Sciences,
University of the West of England, Bristol BS16 1QY, U.K.
Matt-Smith@bigfoot.com, Larry.Bull@uwe.ac.uk

Abstract: The use of machine learning techniques to automatically analyze data for information is becoming increasingly widespread. In this chapter we examine the use of Genetic Programming and a Genetic Algorithm to pre-process data before it is classified using the C4.5 decision tree learning algorithm. Genetic Programming is used to construct new features from those available in the data, a potentially significant process for data mining since it gives consideration to hidden relationships between features. A Genetic Algorithm is used to determine which set of features is the most predictive. Using ten well-known data sets we show that our approach, in comparison to C4.5 alone, provides marked improvement in a number of cases.

3.1 Introduction

Classification is one of the major tasks in data mining, involving the prediction of class value based on information about some other attributes. The process is a form of inductive learning whereby a set of pre-classified training examples are presented to an algorithm which must then generalize from the training set to correctly categorize unseen examples. One of the most commonly used forms of classification technique is the decision tree learning algorithm C4.5 [3.11]. In this chapter we examine the use of Genetic Programming (GP) [3.7] and a Genetic Algorithm (GA) [3.4] to improve the performance of C4.5 through feature *construction* and feature *selection*. Feature construction is a process that aims to discover hidden relationships between features, inferring new composite features. In contrast, feature selection is a process that aims to refine the list of features used thereby removing potential sources of noise and ambiguity. We use GP individuals consisting of a number of separate trees/automatically defined functions (ADFs) [3.7] to construct features for C4.5. A GA is then used to select over the combined set of original and constructed features for a final hybrid C4.5 classifier system. Results show that the system is able to outperform standard C4.5 on a number of data sets held at the UCI repository (http://www.ics.uci.edu/~mlearn/MLRepository.html).

Raymer et al. [3.12] have used ADFs for feature *extraction* in conjunction with the k-nearest-neighbor algorithm. Feature extraction replaces an original feature with the result from passing it through a functional mapping.

In Raymer et al.'s approach each feature is altered by an ADF, evolved for that feature only, with the aim of increasing the separation of pattern classes in the feature space; for problems with n features, individuals consist of n ADFs. Ahluwalia and Bull [3.1] extended Raymer et al.'s approach by coevolving the ADFs for each feature and adding an extra coevolving GA population of feature selectors; extraction and selection occurred simultaneously in $n+1$ populations. For other (early) examples of evolutionary computation approaches to data mining see [3.13] for a GA-based feature selection approach using k-nearest-neighbor and [3.5] for a similar GA-based approach also using k-nearest-neighbor. Since undertaking the work presented here we have become aware of Vafaie and DeJong's [3.14] combination of GP and a GA for use with C4.5. They used the GA to perform feature selection for a face recognition data set where feature subsets were evaluated through their use by C4.5. GP individuals were then evolved which contained a variable number of ADFs to construct new features from the selected subset, again using C4.5. Our approach is very similar to Vafaie and DeJong's but the feature operations are reversed such that feature construction occurs before selection. We find that our approach performs as well or better than Vafaie and DeJong's.

More recent work using GP to construct features for use by C4.5 includes that of Otero et al. [3.10]. Otero et al. use a population of GP trees to evolve a single new feature using information gain as the fitness measure (this is the criteria used by C4.5 to select attributes to test at each node of the decision tree). This produces a single feature that attempts to cover as many instances as possible – a feature that aims to be generally useful and which is appended to the set of original features for use by C4.5. Ekárt and Márkus [3.3] use GP to evolve new features that are useful at specific points in the decision tree by working interactively with C4.5. They do this by invoking a GP algorithm when constructing a new node in the decision tree – e.g. when a leaf node incorrectly classifies some instances. Information gain is again used as the fitness criterion but the GP is trained only on those instances relevant at that node of the tree.

Krawiec [3.8] also uses GP to construct new features for use by C4.5 but instead of using information gain as the fitness criterion uses, like the technique presented here, the 'so-called wrapper approach [3.6] where the evaluation consists of multiple train-and-test experiments carried out [with] the same inducer that is used to create the final classifier'[3.8]. Krawiec justifies the additional computational expense involved by reporting Kohavi and Johns [3.6] findings that 'although computationally demanding ... [the wrapper approach] seems to out-perform other methods on most tasks'. Krawiec's algorithm creates a fixed number of new features (4 in the experiments shown) which *replace* the original set of features without any subsequent selection. Krawiec also extends the algorithm with the concept of features that are hidden from the evolutionary process to preserve them from destruction. These features (2 in the experiments shown) are selected according to the number of times they appear in the decision trees constructed during fitness evaluation. While Krawiec's approach bears some similarity with the algorithm presented here, there are a number of differences: the fixed number of new features introduces a parameter that must be altered for each new problem; it does not involve any subset selection (other than the implicit selection of original features by their presence in the ADFs); nor does it appear to allow for the inclusion of any original features found to be useful.

This contribution is arranged as follows: the next section describes the initial approach. Section 3.3 presents results from its use on a number of well-known data sets and discusses the results. This is followed by some amendments to the algorithm and further results. Finally Section 3.4 presents some conclusions and future directions.

3.2 The GAP Algorithm

In this work we have used the WEKA [3.15] implementation of C4.5, known as J48, to examine the performance of our Genetic Algorithm and Programming (GAP) approach. This is a wrapper approach, in which the fitness of individuals is evaluated by performing 10-fold cross validation using the same inducer as used to create the final classifier: C4.5 (J48). The approach consists of two phases:

3.2.1 Feature Creation

A population of 101 genotypes is created at random. Each genotype consists of n trees, where n is the number of numeric valued features in the data set, subject to a minimum of 7. This minimum is chosen to ensure that, for data sets with a small number of numeric features, the initial population contains a significant number of compound features. A tree can be either an original feature or an ADF. That is, a genotype consists of n GP trees, each of which may contain 1 or more nodes. The chance of a node being a leaf node (a primitive attribute) is determined by:

$$P_{leaf} = 1 - \frac{1}{(depth + 1)} \tag{3.1}$$

where $depth$ is the depth of the tree at the current node. Hence a root node will have a depth of 1, and therefore a probability of 0.5 of being a leaf node. Nodes at depth 2 will have a 0.67 probability of being a leaf node, and so on. If a node is a leaf node, it takes the value of one of the original features chosen at random. Otherwise, a function is randomly chosen from the set {*, /, +, -, %} and two child nodes are generated. In this manner there is no absolute limit placed on the depth any one tree may reach but the average depth is limited. During the initial creation no two trees in a single genotype are allowed to be alike, though this restriction is not enforced in later stages. Additionally, nodes with '–', '%' or '/' for functions cannot have child nodes that are equal to each other. In order to enforce this child nodes within a function '*' or '+' are ordered alphabetically to enable comparison (e.g. [width + length] will become [length + width]).

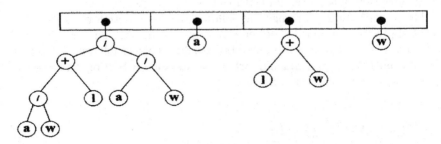

Fig. 3.1. Sample genotype (genotypes have a minimum of 7 trees, but only 4 are shown in the sample genotype due to space constraints. The sample genotype has been constructed using a very simple data set with 3 attributes – area (a), length (l) and width (w))

An individual is evaluated by constructing a new data set with one feature for each tree in the genotype. This data set is then passed to a C4.5 (J48) classifier (using default parameters), whose performance on the data set is evaluated using 10-fold cross validation. The percentage correct is then assigned to the individual and used as the fitness score.

Once the initial population has been evaluated, several generations of selection, crossover, mutation and evaluation are performed. After each evaluation, if the fittest individual in the current generation is fitter than the fittest so far, a copy of it is set aside and the generation noted. The evolutionary process continues until the following conditions are met: at least 10 generations have passed, and the fittest individual so far is at least 6 generations old. There is no maximum generation, but in practice very rarely have more than 50 generations been necessary, and often fewer than 30 are required. This is still a lengthy process, as performing 10-fold cross validation for each member of the population is very processor intensive. The extra time required can justified by the improvement in the results over using, e.g., a single train and test set (results not shown). Information Gain, the fitness criterion employed by both Otero and Ekárt, is much faster but is only applicable to a single feature – it cannot provide the fitness criterion for a set of features.

We use tournament selection to select the parents of the next generation, with tournament size 8 and a 0.3 probability of the fittest individual winning (otherwise a 'winner' is selected at random from the tournament group).

There is a 0.6 probability of two-point crossover occurring between the ADFs of the two selected parents (whole trees are exchanged between genotypes):

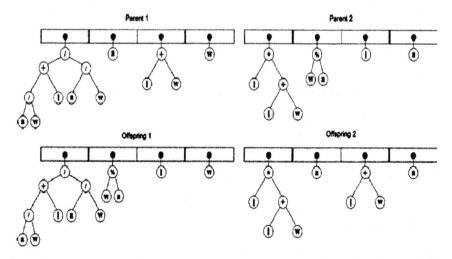

Fig. 3.2. GA crossover

There is an additional 0.6 probability that crossover will occur between two ADFs at a randomly chosen locus (sub-trees are exchanged between trees):

Mutation is used with probability 0.008 per node, whereby a randomly created subtree replaces the subtree under the selected node:

We also use a form of inversion with probability 0.2 whereby the order of the trees between two randomly chosen loci is reversed:

Experimentation on varying these parameters has found the algorithm to be fairly robust to their setting (not shown).

Once the termination criteria have been met the fittest individual is used to seed the feature selection stage.

3.2.2 Feature Selection

The fittest individual from the feature creation stage (ties broken randomly) is analyzed to see if any of the original features do not appear to be used. If there are any missing, sufficient trees are added to ensure that every original feature in the data set appears at least once (the new trees are not randomly generated as in the feature creation stage, but have single nodes containing the required attribute). This extended genotype (up to twice as long as the fittest individual of the feature creation stage) replaces the initial individual and is used as the basis of the second stage.

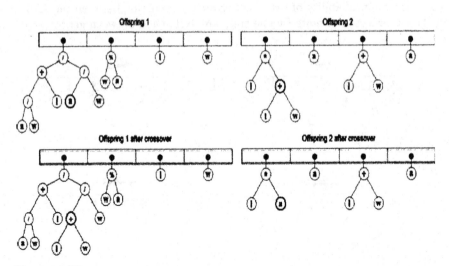

Fig. 3.3. GP crossover

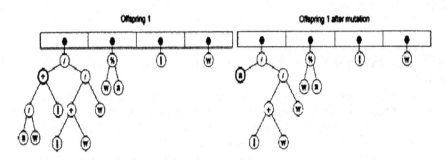

Fig. 3.4. Mutation

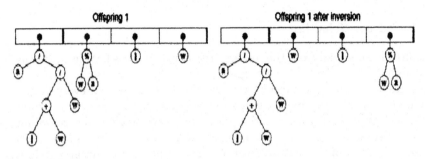

Fig. 3.5. Inversion

A new data set is constructed with one attribute for every tree in the extended genotype. In an attempt to reduce over-fitting of the data, the order of the data set is randomly reordered at this point. This has the effect of providing a different split of the data for 10-fold cross validation during the selection stage, giving the algorithm a chance of recognizing trees that performed well only due to the particular data partition in the creation stage. As a result of the reordering, it is usually the case that the fitness score of individuals in the selection stage is less than that of individuals in the creation stage, but solutions should be more robust.

For the GA a population of 101 bit strings is randomly created. The strings have the same number of bits as the genotype has trees – there is one bit for every attribute (some composite, some primitive attributes). The last member of the population, the 101st bit string, is not randomly created but is initialized to all ones. This ensures that there are no missing alleles at the start of the selection.

Once the entire population has been created, each genotype is evaluated and assigned a fitness score that is used in selecting the parents of the next generation. A GA bit string is evaluated by taking a copy of the parent data set and removing every attribute that has a '0' in the corresponding position in the bit string. As in the feature creation stage, this data set is then passed to a C4.5 (J48) classifier whose performance on the data set is evaluated using 10-fold cross validation. The percentage correct is then assigned to the bit string and used as the fitness score.

If the fittest individual in the current generation has a higher fitness score than the fittest so far (from the selection stage, ignoring genotypes from the feature creation stage), or it has the same fitness score and fewer '1's, a copy of it is set aside and the generation noted. As in the feature creation stage the cycles of selection, crossover, mutation, and evaluation continue until the following conditions are met: at least 10 generations have passed, and the fittest individual so far is at least 6 generations old.

The selection scheme is the same as for the creation stage. There is a 0.6 probability of two-point crossover and a 0.005 per bit probability of mutation.

3.3 Experimentation

We have used ten well-known data sets from the UCI repository to examine the performance of the GAP algorithm. The UCI data sets were chosen because they consisted entirely of numeric attributes (though the algorithm can handle some nominal attributes, as long as there are two or more numeric attributes present) and had no missing values (missing values could be handled with only a minor modification to the code). Table 3.1 shows the details of the ten data sets used here.

For performance comparisons the tests were performed using ten-fold cross-validation (in which 90% of the data was used for training and 10%

Table 3.1. UCI data set information

Data set	Numeric features	Nominal features	Classes	Instances
BUPA Liver Disorder (Liver)	6	0	2	345
Glass Identification (Glass)	9	0	6	214
Ionosphere (Iono.)	34	0	2	351
New Thyroid (NT)	5	0	3	215
Pima Indians Diabetes (Diab.)	8	0	2	768
Sonar	60	0	2	208
Vehicle	18	0	4	846
Wine Recognition (Wine)	13	0	3	178
Wisconsin Breast Cancer – New (WBC New)	30	0	2	569
Wisconsin Breast Cancer – Original (WBC Orig.)	9	0	2	699

for testing). An additional set of ten runs using ten-fold cross validation were made (a total of twenty runs - two sets of ten-fold cross-validation) to allow a paired t-test to establish the significance of any improvement over C4.5 (J48).

3.3.1 Results

The highest classification score for each data set is shown in Table 3.2 in bold underline. The first two columns show the performance of the GAP algorithm on the training data and the last column shows the results of the paired t-test. Results that are significant at the 95% confidence level are shown in bold.

The GAP algorithm out-performs C4.5 (J48) on eight out of ten data sets, and provides a significant improvement on three (Glass Identification, New Thyroid, and Wisconsin Breast Cancer Original) – two of which are significant at the 99% confidence level. There are no data sets on which the GAP algorithm performs significantly worse than C4.5(J48) alone.

The standard deviation of the GAP algorithm's results do not seem to differ greatly from that of C4.5 (J48); there are five data sets where the GAP algorithms' standard deviation is greater and five where it is smaller. This is perhaps the most surprising aspect of the results, given that the GAP algorithm is unlikely to produce the same result twice when presented with exactly the same data, whereas C4.5 (J48) will always give the same result if presented with the same data.

As noted in the introduction, Vafaie and DeJong [3.14] have used a very similar approach to improve the performance of C4.5. They use feature selection (GA) followed by feature construction (GP). We have examined the performance of our algorithm as described above with the two processes occurring in the opposite order. Results indicate that GAP gives either equivalent (e.g. Wisconsin Breast Cancer) or better performance (e.g. New Thyroid)

Table 3.2. Comparative performance of GAP algorithm and C4.5 (J48)

Data set	GAP Train	S.D.	GAP Test	S.D.	C4.5 (J48)	S.D.	Paired t-test
Liver	75.04	2.37	65.97	11.27	**66.37**	8.86	-0.22
Glass	78.17	1.95	**73.74**	9.86	68.28	8.86	**3.39**
Iono.	95.99	0.87	89.38	4.76	**89.82**	4.79	-0.34
NT	98.22	0.68	**96.27**	4.17	92.31	4.14	**3.02**
Diab.	78.15	0.96	**73.50**	4.23	73.32	5.25	0.19
Sonar	92.33	1.27	**73.98**	11.29	73.86	10.92	0.05
Vehicle	78.82	1.21	**72.46**	4.72	72.22	3.33	0.20
Wine	98.47	0.80	**94.68**	5.66	93.27	5.70	0.85
WBC New	97.86	0.47	**95.62**	2.89	93.88	4.22	1.87
WBC Orig.	97.65	0.38	**95.63**	1.58	94.42	3.05	**2.11**
Overall Average	89.07		83.12		81.77		2.91

(Table 3.3). We suggest this is due to GAP's potential to construct new features in a less restricted way, i.e. its ability to use all the original features during the create stage. For instance, on the New Thyroid data set the select stage will always remove either feature a or feature b thus preventing the create stage from being able to construct the apparently useful feature "b/a" (see Section 3.3.2). That is, on a number of data sets there is a significant difference (at the 95% confidence level) in the results brought about by changing the order of the stages.

Table 3.3. Comparison of ordering of create and select stages

Dataset	Create then Select Test	S.D.	Select then Create Test	S.D.
Liver	65.97	11.27	**67.42**	8.23
Glass	**73.74**	9.86	68.75	6.36
Iono.	**89.38**	4.76	89.02	4.62
NT	**96.27**	4.17	93.67	5.82
Diab.	73.50	4.23	**74.09**	4.46
Sonar	**73.98**	11.29	73.16	9.05
Vehicle	72.46	4.72	72.46	3.01
Wine	94.68	5.66	**94.69**	3.80
WBC New	95.62	2.89	**95.88**	3.15
WBC Orig.	**95.63**	1.58	95.13	2.16
Overall Average	83.12		82.43	

3.3.2 Analysis

We were interested in whether the improvement over C4.5 (J48) is simply the result of the selection stage choosing an improved subset of features and discarding the new constructed features. An analysis of the attributes output by the GAP classifier algorithm, and the use made of them in C4.5's decision trees, shows this is not the case.

As noted above, the results in Table 3.2 were obtained from twenty runs on each of ten UCI data sets, i.e. a total of two hundred individual solutions. In those two hundred individuals there are a total of 2,425 trees: 982 ADFs and 1,443 original features - a ratio of roughly two constructed features to three original features. *All but two of the two hundred individuals contained at least one constructed feature.* Table 3.4 gives details of the average number of ADFs per individual for each data set (the number of original features used is not shown).

Table 3.4. Analysis of the constructed features for each data set

Data set		Results		
Name	Features	Average Features	Average ADFs	Minimum ADFs
Liver	6	4.9	2.6	1
Glass	9	7.1	3.1	1
Iono.	34	19.7	7.6	4
NT	5	3.2	2.3	1
Diab.	8	6.4	2.7	1
Sonar	60	38.0	13.6	5
Vehicle	18	16.3	5.5	2
Wine	13	5.2	2.1	0
WBC New	30	15.7	7.0	4
WBC Orig.	9	5.0	2.9	1

Knowing that the feature selection stage continues as long as there is a reduction in the number of attributes without reducing the fitness score, we can assume that C4.5 (J48) is making good use of all the attributes in most if not all of the winning individuals. This can be demonstrated by looking in detail at the attributes in a single winner, and the decision tree created by C4.5(J48).

One of the best performers on the New Thyroid data set had three trees, two of them ADFs and hence constructed features. The original data set contains five features and the class:

1. T3-resin uptake test. (A percentage)
2. Total Serum thyroxin as measured by the isotopic displacement method.
3. Total serum triiodothyronine as measured by radioimmuno assay.
4. Basal thyroid-stimulating hormone (TSH) as measured by radioimmuno assay.
5. Maximal absolute difference of TSH value after injection of 200 micro grams of thyrotropin-releasing hormone as compared to the basal value.

Class attribute. (1 = normal, 2 = hyper, 3 = hypo)
In the chosen example the newly constructed features were:

- "e" becomes Feat0
- "((a/d)*b)" becomes Feat1
- "(b/a)" becomes Feat2

The decision tree created by C4.5 (the numbers after the class prediction indicate the count of train instances correctly / incorrectly classified) was:

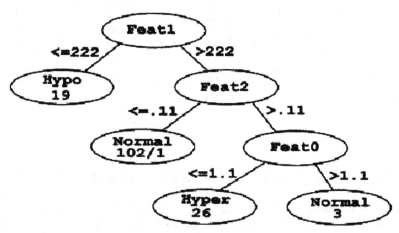

Fig. 3.6. New thyroid decision tree

It is apparent that C4.5 (J48) is using the constructed features to classify a large majority of the instances, and only referring to one of the original features (Feat0, or the original feature "e") in 29 of 150 cases.

3.3.3 The Importance of Reordering the Dataset

In section two it was mentioned that the data set was randomly reordered before the second stage commenced, providing a different view of the data for 10-fold cross validation during fitness evaluation and, it was hoped, this would reduce over fitting and improve the performance on the test data. Is this what actually happens? In order to test this hypothesis we turned off the randomization and retested the algorithm. The first impression is that there is no important difference between the two sets of results – there are 5 data sets where not reordering gives a better result and 5 where it is worse. However, there are now only two (rather than three) data sets on which the algorithm provides a significant improvement over C4.5 (J48) (New Thyroid and Wisconsin Breast Cancer); and most importantly the t-test performed over the 200 runs from all data sets no longer shows a significant improvement. The results were as follows (the column for paired t-test shows the results for testing the algorithm without reordering against C4.5(J48)):

Table 3.5. Comparative performance of GAP algorithm with and without reordering) and C4.5 (J48)

Data set	GAP reorder	S.D.	GAP no reorder	S.D.	C4.5 (J48)	S.D.	Paired t-test
Liver	65.97	11.27	**66.65**	7.84	66.37	8.86	0.17
Glass	**73.74**	9.86	69.74	9.79	68.28	8.86	0.61
Iono.	89.38	4.76	89.77	4.24	**89.82**	4.79	-0.04
NT	96.27	4.17	**97.22**	3.78	92.31	4.14	**3.99**
Diab.	**73.50**	4.23	71.74	4.34	73.32	5.25	-1.37
Sonar	73.98	11.29	**75.22**	8.32	73.86	10.92	0.50
Vehicle	**72.46**	4.72	71.94	4.43	72.22	3.33	-0.28
Wine	**94.68**	5.66	94.08	5.09	93.27	5.70	0.55
WBC New	**95.62**	2.89	94.56	2.58	93.88	4.22	0.71
WBC Orig.	95.63	1.58	**95.71**	1.59	94.42	3.05	**2.03**
Overall Average	83.12		82.66		81.77		1.82

3.3.4 Combining Creation and Selection in a Single Stage

Having successfully tested the algorithm with two separate stages, we redesigned it to move feature selection into the construction stage. Feature construction occurs as before but each tree now has a bit flag associated with it, to determine whether the tree is passed to C4.5(J48) for evaluation. During crossover each tree retains its associated bit flag, which is subject to the same chance of mutation as during the second stage (0.005).

Testing the amended algorithm with the same parameter values as before gives a much shorter run time (not surprisingly, roughly half the time of the two stage algorithm) but with poorer results – an overall average of 82.20%[1] (though this is still an improvement over unaided C4.5 (J48)).

There are three primary differences between the two versions of the algorithm that may account for this drop in performance:

1. With a single stage we are asking the algorithm to do the same amount of work in half the time.
2. There is no longer an opportunity to randomly reorder the data set between stages.
3. There is no longer an opportunity to reintroduce any original attributes that have been dropped during the first stage.

There seems no reasonable way to address the third of these differences with a single stage approach, but the other two can be compensated for. Firstly we

---

[1] It should be noted that the results for some of the data sets have a fairly high standard deviation, and so can show some variation in the results from run to run. For this reason we have taken to using the average result over all 10 data sets as a useful (and briefer!) indicator of the performance of the algorithm.

can change the termination criteria – by doubling both the minimum number of generations to 20 and the age of the fittest individual to 12 generations. Doing this does improve the result (to an overall average result of 82.88%) but not sufficiently to bring it into line with a two stage process.

Additionally we can randomly reorder the data set. We considered two approaches to this. The first was to have two versions of the data set from the start, with the same data but in a different order, and simply alternate between data sets when evaluating each generation (i.e. the first data set was used to evaluate even numbered generations, the second to evaluate odd numbered) – this approach did not seem to improve the results (slightly worse than having no reordering at 82.24%). The second, more successful, approach was to reorder the data set once the termination criteria had been reached. That is, run as before but when the fittest individual reaches 12 generations old reorder the data set, re-evaluate the current generation and reset the fittest individual, then continue until the termination criteria are met again. The results obtained with a longer run time and randomly reordering the data set part-way through are shown in the table below (Table 3.6) (the column for paired *t*-test shows the results for testing the single stage algorithm against C4.5 (J48)).

Table 3.6. Comparative performance single stage and C4.5 (J48)

Data set	2 stage	S.D.	1 stage	S.D.	C4.5 (J48)	S.D.	Paired t-test
Liver	65.97	11.27	**66.55**	8.10	66.37	8.86	0.11
Glass	**73.74**	9.86	71.84	10.26	68.28	8.86	1.78
Iono.	89.38	4.76	**90.69**	4.66	89.82	4.79	0.96
NT	96.27	4.17	**96.49**	3.98	92.31	4.14	**3.69**
Diab.	73.50	4.23	**73.64**	5.11	73.32	5.25	0.24
Sonar	73.98	11.29	**75.89**	9.00	73.86	10.92	0.80
Vehicle	**72.46**	4.72	72.11	4.60	72.22	3.33	-0.09
Wine	94.68	5.66	**96.10**	4.08	93.27	5.70	1.76
WBC New	95.62	2.89	**95.71**	4.39	93.88	4.22	1.71
WBC Orig.	**95.63**	1.58	95.56	2.52	94.42	3.05	1.62
Overall Average	83.12		83.46		81.77		3.62

Although the single stage algorithm out-performs C4.5 (J48)) on only one data set at the 95% confidence level (a *t*-test of 1.96 or higher), as compared to three data sets for the two stage algorithm, it outperforms C4.5(J48) on everything but the Vehicle data set (and then loses by a very small margin). It also improves on the performance of the two stage version on seven out of ten data sets, resulting in an increase of the (already high) overall confidence of improvement over C4.5 (J48).

3.3.5 A Rough Comparison to Other Algorithms

Table 3.7 presents a number of published results we have found regarding the same ten UCI data sets using other machine learning algorithms. It should be noted that in the table the results for C4.5 were not obtained using J48. The GAP column presents results for the single stage version of the algorithm. Cells in the table are left blank where algorithms were not tested on the data set in question. The highest classification score for each data set is shown in bold underline.

Table 3.7. Performance of GAP and other algorithms on the UCI data sets

Data set	GAP	C4.5 (J48)	C4.5	HIDER	XCS	O. F. A.	LVSM
Liver	66.55	66.37	65.27	64.29	67.85	57.01	**68.68**
Glass	**71.84**	68.28	67.27	70.59	72.53	69.56	
Iono.	**90.69**	89.82					87.75
NT	**96.49**	92.31					
Diab.	73.64	73.32	67.94	74.1	68.62	69.8	**78.12**
Sonar	75.89	73.86	69.69	56.93	53.41	**79.96**	
Vehicle	72.11	72.22					
Wine	96.10	93.27	93.29	96.05	92.74	**98.27**	
WBC New	**95.71**	93.88					
WBC Orig.	95.56	94.42	93.72	95.71	**96.27**	94.39	

The results for C4.5, HIDER and XCS were obtained from [3.2], those for O.F.A. ('Ordered Fuzzy ARTMAP', a neural network algorithm) from [3.1] and LVSM (Lagrangian Support Vector Machines) from [3.9].

The differences between the reported results for C4.5 and those for C4.5 as used in this paper (J48, the WEKA implementation of C4.5) are likely to arise from different data partitions used for the tests (the most notable being the 5.38% difference in results for Pima Indians Diabetes). This discrepancy highlights the dangers inherent in comparing results with published data – the comparison should be seen as purely informal. The only comparisons that can be relied upon are those between the GAP classifier and C4.5 (J48) as these have been performed using exactly the same procedure and data partitions. The results are by no means an exhaustive list of current machine learning algorithms, nor are they guaranteed to be the best performing algorithms available, but they give some indication of the relative performance of our approach – which appears to be very good.

3.4 Conclusion

In this chapter we have presented an approach to improve the classification performance of the well-known induction algorithm C4.5. We have shown that

GP individuals consisting of multiple trees/ADFs can be used for effective feature creation and that solutions, combined with feature selection via a GA in either a separate or the same stage, can give significant improvements to the classification accuracy of C4.5. We have also indicated that randomly reordering the data set part-way through the process may help to reduce the problem of overfitting.

Future work will apply our approach to other data sets and data mining algorithms.

References

3.1 Ahluwalia, M., Bull, L. (1999): Co-evolving functions in genetic programming: classification using k-nearest neighbour. In Banzhaf, W., Daida, J., Eiben, G., Garzon, M. H., Honavar, J., Jakeila, K., Smith, R. (eds) GECCO-99: Proceedings of the Genetic and Evolutionary Computation Conference. Morgan Kaufmann, 947–952

3.2 Dixon, P. W., Corne, D. W., Oates, M. J. (2001): A preliminary investigation of modified XCS as a generic data mining tool. In Lanzi, P. L., Stolzmann, W., Wilson, S. (eds) Advances in Learning Classifier Systems. Springer, pp.133-151

3.3 Ekárt, A., Márkus, A. (2003): Using genetic programming and decision trees for generating structural descriptions of four bar mechanisms. To appear in Artificial Intelligence for Engineering Design, Analysis and Manufacturing

3.4 Holland, J.H. (1975): Adaptation in natural and artificial systems. Univ. Michigan

3.5 Kelly, J.D., Davis, L. (1991): Hybridizing the genetic algorithm and the k nearest neighbors classification algorithm. In R. Belew, L. Booker (eds) Proceedings of the Fourth International Conference on Genetic Algorithms. Morgan Kaufmann, pp377-383

3.6 Kohavi, R., John, G. H. (1997): Wrappers for feature subset selection. Artificial Intelligence Journal **1,2**, 273-324

3.7 Koza, J.R. (1992): Genetic programming. MIT Press

3.8 Krawiec, Krzysztof (2002): Genetic programming-based construction of features for machine learning and knowledge discovery tasks. Genetic Programming and Evolvable Machines 3, 329-343

3.9 Mangasarian, O. L., Musicant, D. R. (2001): Lagrangian support vector machines. Journal of Machine Learning Research, **1**, 161-177

3.10 Otero, F. E. B., Silva, M. M. S., Freitas, A. A., Nievola, J. C. (2003): Genetic programming for attribute construction in data mining. In Ryan, C., Soule, T., Keijzer, M., Tsang, E., Poli, R., Costa, E. (Eds.) Genetic Programming: 6th European Conference, EuroGP 2003, Essex, UK, Proceedings. Springer, 384-393

3.11 Quinlan, J.R. (1993): C4.5: Programs for machine learning. Morgan Kaufmann

3.12 Raymer, M.L., Punch, W., Goodman, E.D., Kuhn, L. (1996): Genetic programming for improved data mining - application to the biochemistry of protein interactions. In Koza, J. R., Deb, K., Dorigo, M., Fogel, D. B., Garzon, M., Iba, H., Riolo, R. (eds) Proceedings of the Second Annual Conference on Genetic Programming, Morgan Kaufmann, 375-380

3.13 Siedlecki, W., Sklansky, J. (1988): On automatic feature selection. International Journal of Pattern Recognition and Artificial Intelligence 2, 197-220

3.14 Vafaie, H., De Jong, K. (1995): Genetic algorithms as a tool for restructuring feature space representations. In Proceedings of the International Conference on Tools with A.I. IEEE Computer Society Press

3.15 Witten, I.H., Frank, E. (2000): Data mining: practical machine learning tools and techniques with Java implementations. Morgan Kaufmann

4. Multi-Agent Data Mining using Evolutionary Computing

Riyaz Sikora

Dept. of Information Systems & OM
The University of Texas at Arlington
P.O. Box 19437, Arlington, TX 76019
rsikora@uta.edu

Abstract : In this chapter we present a multi-agent based data mining algorithm for scaling up and improving the performance of traditional inductive learning algorithms that build feature-vector-based classifiers in the form of rule sets. With the tremendous explosion in the amount of data being amassed by organizations of today, it is critically important that data mining techniques are able to process such data efficiently. We present the Distributed Learning System, a multi-agent based data mining system that uses genetic algorithms as learning agents and incorporates a self-adaptive feature selection method.

4.1 Introduction

The amount of customer, financial, marketing, operations, and other sorts of data being amassed by organizations has increased by manifold in recent times. The ability to produce and store such voluminous data has far outpaced the ability to analyze and interpret this data, and derive useful knowledge from it. Large databases these days have millions of records and each record may have hundreds or even thousands of fields. For example, in the business world, one of the largest databases was created by Wal-Mart, which handles over 20 million transactions a day [4.3]. Similar instances can be found in databases created by health care companies, oil exploration firms, database marketing, and scientific research consortiums, just to name a few.

Such volumes of data clearly overwhelm more traditional data analysis methods. A new generation of tools and techniques are needed to find interesting patterns in the data and discover useful knowledge. Successful development of effective and efficient data mining algorithms can also provide enormous benefits to an organization from the business standpoint. Benefits include reduced costs due to more accurate control, more accurate future predictions, more effective fault detection and prediction, fraud detection and control, and automation of repetitive human tasks.

Although there are several algorithms available for discerning patterns from data in the machine learning, statistics, and classification literature, they generally break down in data mining applications because of the size of the data set. In a typical application involving data mining there are hundreds, possibly thousands of fields per record, majority of which are either unimportant or not directly relevant to the problem. The traditional machine learning algorithms end up wasting a lot of computational effort processing unimportant information.

In this chapter we present the concepts of distributed learning and simultaneous feature selection in designing more effective and efficient genetic algorithm (GA) based data mining algorithms. The distributed learning concept uses the divide-and-conquer strategy in which the data set is divided into multiple sub-sets. Multiple copies of the algorithm then work on the sub-sets in parallel and their results are synthesized. This concept is especially suitable for evolutionary computation methods, as they are population based. The results of the sub-sets can be used as members of the population in the synthesis step. Because of the parallelism involved in distributed learning, the developed algorithm can also be executed more efficiently on a parallel machine.

4.2 Related Work

There are many data mining algorithms currently in use. Most of them can be classified in one of the following categories: Decision trees and rules [4.10, 4.37], nonlinear regression and classification methods [4.19, 4.16], example-based methods [4.13, 4.23], probabilistic graphical dependency models [4.32, 4.47], and relational learning models [4.14]. Over the years genetic algorithms have been successfully applied in learning tasks in different domains, like chemical process control [4.40], financial classification [4.41], manufacturing scheduling [4.28], robot control [4.42], etc. There has also been some work done related to developing hybrid learning systems involving genetic algorithms [4.5, 4.46].

One of the biggest problems in using some of the traditional machine learning methods for data mining is the problem of scaling up the methods to handle the huge size of the data sets and their high-dimensionality. Provost and Kolluri [4.35] provide a survey of techniques for scaling up machine learning algorithms. In this chapter we look at two approaches, feature selection and distributed learning, that can be used for making learning algorithms more efficient and scalable.

4.2.1 Feature Selection

Feature selection can be defined as selecting the smallest subset of the original set of features that are necessary and sufficient for describing the target concept [4.24]. The marginal benefit resulting from the presence of a feature in a given set plays an important role. A given feature might provide more information when present with certain other feature(s) than when considered by itself. Cover [4.12], Elashoff *et. al.* [4.15], and Toussaint [4.44], among others, have shown the importance of selecting features as a set, rather than selecting the best features to form the (supposedly) best set. They have shown that the best individual features do not necessarily constitute the best set of features.

There have been many approaches to feature selection based on a variety of techniques, such as statistical [4.25], geometrical [4.17], information-theoretic measures [4.7], mathematical programming [4.9], among others. Several researchers have also used evolutionary algorithms for feature selection by using a classifier as a fitness function [4.36, 4.6, 4.46, 4.45]. GAs have also been used in feature selection for creation of ensemble classifiers [4.20, 4.31].

Feature selection has been traditionally used in data mining applications as part of the data cleaning and/or pre-processing step where the actual extraction and learning of knowledge or patterns is done after a suitable set of features is extracted. If the feature selection is independent of the learning algorithm it is said to use a *filter* approach. Lanzi [4.27] presents one such filter approach to feature selection using genetic algorithms.

If the feature selection method works in conjunction with the learning algorithm it is using a *wrapper* approach [4.22, 4.48]. The problem with the filter approach is that the optimal set of features may not be independent of the learning algorithm or classifier. The wrapper approach, on the other hand, is computationally expensive as each candidate feature subset has to be evaluated by executing a learning algorithm on that subset.

When used with GAs, the wrapper approaches become even more prohibitively expensive [4.27]. Raymer *et. al.* [4.38] presents an approach of simultaneously doing feature selection and optimizing feature weights using a GA. In this chapter we present a new approach to feature selection using genetic algorithms that is based on the concept of self-adaptation where the learning and feature selection are done simultaneously.

Feature selection concept presented in this chapter is related to two aspects of work done earlier: self-adaptation and use of non-coding material in the chromosome structure (called *introns*) motivated by the existence of non-encoding DNA in biological systems. In biological systems an intron is a portion of the DNA that is not transcribed into proteins. Introns can become an important part of the evolution process by providing a buffer against the destructive effects of the genetic algorithm. At the same time introns have been shown to be useful as a source of symbols that can be effectively used to evolve new behaviors through subsequent evolution [4.29, 4.30].

Self-adaptation [4.1] refers to the technique of allowing characteristics of the search to evolve during the search rather than be specified by the user. Most of the work done on self-adaptation has focused on choices related to search operators. Aspects of these choices are encoded along with each member of the population and they are allowed to vary and adapt on an individual basis. One of the most common traits to be self-adapted has been the mutation rate [4.39, 4.18, 4.4]). Others have included crossover [4.2] and inversion operators [4.11].

In all applications involving the use of evolutionary computation methods in machine learning, feature selection is either done implicitly or a wrapper approach is used for discerning important set of features. The learning algorithm is then applied for learning rules or patterns based on the features selected. In this chapter we present a GA for doing both of these tasks simultaneously by evolving a binary code for feature selection along side the chromosome structure used for evolving the rules.

4.2.2 Distributed Data Mining

Another approach for improving the efficiency of data mining algorithms is by using distributed learning. Even if all the data is centrally located in a single database, distributed data mining can be used effectively in scaling up by using the divide-and-conquer strategy. Moreover, there are many data mining problems that are inherently distributed. For example, the point-of-sale information for any big retailer (like Wal-Mart) is distributed across its various locations. In many instances it is not feasible to create and maintain a monolithic database by combining these distributed sources into a centralized database. In other scenarios security concerns do not allow the coalescing of separate data sources into one database. In such cases distributed data mining is the only feasible alternative, whereby the local databases are mined independently by the data mining algorithms and the results are then combined.

Provost [4.34] provides a comprehensive summary of the different factors that motivate the development of distributed data mining. Prodomidis [4.33] present an approach for distributed data mining by using Java agents in which the distributed databases are mined independently by separate learning agents and their results are combined by using a meta-learning strategy. In this chapter we present a multi-agent system for distributed data mining that uses GA-based learning algorithm as individual agents and show that it can significantly improve the scale-up properties of the learning agent.

4.3 Data Mining with GA

In this section we present the design of a genetic algorithm for rule learning in a data mining application. Assume that the data mining problem has k attributes and we have a set of training examples

$\psi \quad = \{(E_t, c) \mid t = 1, \ldots, T\}$

where T is the total number of examples, each example E_t is a vector of k attribute values

$E_t = [e_{t1}, e_{t2}, \ldots, e_{tk}]$

and c is its classification value (usually a binary value indicating whether the example is a positive or a negative). The goal of data mining is to learn concepts that can explain or cover all of the positive examples without covering the negative examples.

The representation of a concept or a classifier used by the GA is that of a disjunctive normal form. A concept is represented as

$$\Omega = \delta_1 \vee \delta_2 \vee \ldots \vee \delta_p$$

where each disjunct δ_i (also referred to as a rule) is a conjunction of conditions on the k attributes,

$$\delta_i = (\xi_{1,i} \wedge \ldots \wedge \xi_{k,i}).$$

The above concept Ω is said to have a size of p (which we refer to as the rule size). In order to handle *continuous attributes* each condition $\xi_{j,i}$ is in the form of a closed interval $[a_j, b_j]$. We say that a disjunct δ_i covers an example E_t if

$$(a_j \leq e_{tj} \leq b_j) \forall j = 1 \ldots k.$$

Each member of the population in the GA is a single disjunct and the GA tries to find the best possible disjunct. At each generation it retains the best disjunct and replaces the rest through the application of the genetic operators. After the genetic algorithm converges, the best disjunct found is retained and the positive examples it covers are removed. The process is repeated until all the positive instances are covered. The final rule or concept is then the disjunct of all the disjuncts found. This procedure for searching the instance space (I-space) is called *explanation based filtering*, and is summarized in Fig. 4.1.

The fitness function looks at the number of positive and negative examples covered by the rule but it also assigns partial credit for the number of attribute intervals on that rule that match the corresponding attribute values on a positive training example. Specifically, the fitness function is given by

$$F = \alpha + Ck(p - n),$$

where α is the total number of partial matches, C is a constant, k is the number of attributes, p is the number of positive examples covered by the rule, and n is the number of negative examples covered by the rule. Note that the partial credit portion of the fitness functions plays an important role in the initial generations in guiding the concepts that are being developed towards covering positive examples. As the concepts start covering more and more positive examples, the second term in the fitness function becomes correspondingly more dominant. The fitness function thus behaves like an adaptive function. In the next section we present a technique for performing simultaneous feature selection using the GA.

```
Input: set of training examples ψ
Begin:
initialize concept description Ω = ∅
while (there are still positive examples in ψ )
{
    initialize the GA with random disjuncts {δᵢ | i = 1 … N} where N is
                                            the population size;
    repeat
        compute F(δᵢ), the fitness function value for each disjunct δᵢ;
        reproduce a new population by applying the genetic operators;
    until ( the stopping criteria for the GA is met );
    Ω = Ω ∨ δ_best (add the best disjunct to the concept);
    remove all the positive examples from ψ that are covered by δ_best;
}
Output: concept description Ω learned by the GA
```

Fig. 4.1. Procedure GA

4.4 Feature Selection as Self-Adaptation

The GA presented above is modified as follows to incorporate the self-adaptive feature selection method. Each population member contains, in addition to the disjunct, a binary vector for feature selection that also evolves alongside the disjunct. A feature is selected if the corresponding bit in the selection code is 1. For example, if we have five attributes in the original feature set a typical rule represented in the new approach would look like the following:

X_1	X_2	X_3	X_4	X_5
((0.0, 0.10)	(0.12, 0.24)	(0.23, 0.50)	(0.4, 0.70)	(0.2, 0.87))
0	1	0	1	1

Since only X_2, X_4, and X_5 have a corresponding 1 on the selection vector, the rule becomes:

IF $(0.12 \leq X_2 \leq 0.24)$ AND $(0.4 \leq X_4 \leq 0.7)$ AND $(0.2 \leq X_5 \leq 0.87)$

The uniform crossover operator is applied to the trio of values (the interval limits and the selection bit) for each attribute instead of the pair of values as in the last section. The mutation operator flips the binary digits on the selection vector in addition to changing the pair of numbers for each attribute as in the last section. The initial population is created as before with the interval pairs for each attribute on a member created with uniform distribution and the binary selection vector randomly generated with a probability of *Fselect* for selecting a particular feature (i.e., a 1 appearing on the selection vector corresponding to that feature). The fitness function remains the same as before except that the selection vector is also used in deciding which attributes are considered for fitness evaluation. Since the fitness evaluation now also depends on the feature subset selected we can hypothesize that this would start evolutionary pressures for good features to be selected.

Note that an attribute's interval limits on a rule can change due to crossover or mutation even when that feature is not selected in the rule. This is similar to the concept of *introns* mentioned earlier where non-coding genes are allowed to propagate and evolve in the hope that in a later generation they might be found to be useful. The interval limits of the attributes not selected in a rule can be thought of as introns. Since these are not used in the fitness evaluation they do not affect the computation time. The only additional resource they consume is the memory storage. However, since the GA used is a fixed length GA the use of introns in this case does not lead to the problems of bloat so often associated with the use of introns in the genetic programming community.

4.5 Multi-Agent Data Mining

The task of data mining is concerned with deriving rules or patterns that can best explain a set of data points or examples. In the traditional approach to data mining, a single learning program is used that generates a set of rules or patterns and successively refines it to explain all the examples. Since the process involves generating and evaluating hypotheses at each stage, it can be more effective and efficient to use a multi-agent approach where the examples are distributed to different learning agents and their partial results are synthesized into the final hypothesis.

Such a multi-agent problem-solving approach is implemented in the Distributed Learning System (DLS). Since a learning agent is now working on only a fraction of the original data set and the different agents can work asynchronously, this method of distributing the amount of resources (data) can make the process parallel. At the same time, by using multiple agents that provide several different hypotheses of the solution, the approach can potentially provide better performance.

The Multi-Agent Systems (MAS) solving paradigm has been extensively studied in the Distributed Artificial Intelligence (DAI) [4.21, 4.8] community. We designed the DLS using the four basic steps of distributed problem solving (Smith and Davis, 1981): (1) problem decomposition, (2) sub problem allocation, (3) sub problem solution, and (4) solution synthesis. Fig. 4.2 illustrates the conceptual model of DLS. At first, the data set P is decomposed into different subsets P_1, $P_2...P_n$, which are then allocated to different learning agents. Each agent solves its sub problem independently of the other agents. The individual solutions are then synthesized into a final solution.

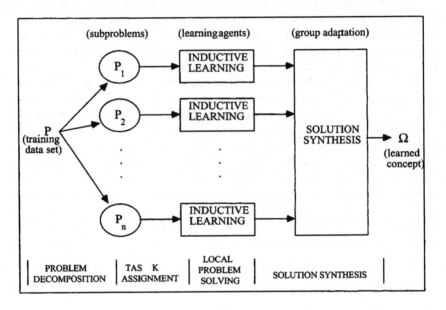

Fig. 4.2. The distributed learning system

The above is only a conceptual model of the DLS and it can be implemented in many different ways. There could be several ways of performing the problem decomposition. For example, in *record-based decomposition* the data set is decomposed along the records. Each learning agent gets a fraction of the records (or examples) from the data set. This models real world situations where the records are distributed. For example, a chain store collecting point-of-sale information at its various locations that are geographically dispersed. A second method of decomposing the data set is the *attribute-based decomposition* where the data set is distributed among the agents along the attributes, with each agent getting the data corresponding to a subset of the attributes. This models the real world situations in which the data is spatially distributed (for e.g., air traffic control). The above mentioned distributed approach of data mining can be very efficient in these cases as the data can be analyzed at its point of origin and the individual results can then be synthesized. The two types of decomposition strategies can also be combined together.

The task assignment and local problem-solving phase of the DLS can also be implemented in several different ways. For example, one can either use the same learning agent on the different subsets or use different types of learning agents. By using different learning agents it is possible to build a hybrid data mining system that can use one learning agent's strengths to mask the problems of others and at the same time provide a diversity of learning biases that are inherent in any learning agent.

A genetic algorithm becomes the natural choice for the synthesis phase of the DLS because of several reasons. First, since it is a population based method it can be seeded with the results of different learning agents. Second, since the GA works by combining two or more solutions to produce better ones it is a perfect choice for synthesizing and improving multiple solutions. Lastly, since its fitness function can be user defined it is ideal for managing trade-offs such as rule size vs. accuracy. Note that for handling complex representation languages one can also use a Genetic Programming [4.26] based synthesizing agent in the DLS that would treat the output of the learning agents as programs. Also, in our version of the DLS the agents do not cooperate explicitly with each other. We use the GA to simulate the cooperative group-problem solving behavior of agents where the group iteratively goes through modifying their individual solutions until they reach some kind of a group solution.

In the DLS system, the data set is distributed over the learning agents. In our implementation all the learning agents use the GA procedure described in Fig. 4.1. Rules learned by the individual agents are then synthesized by another GA to produce the final concept. The DLS procedure is summarized in Fig. 4.3.

Input: a set of training examples ψ,

a group of n data mining agents, π_i, $i = 1 \ldots n$;

Begin:

decompose ψ into n subsets, ψ_1, ψ_2, ... ψ_n, where $\psi = \psi_1 \cup \psi_2 \ldots \psi_n$;

for each agent π_i, $i = 1 \ldots n$, execute the procedure GA(ψ_i) to get

$$\Omega_i;$$

create a population Γ by using the individual disjuncts from all

the concepts Ω_i, $i = 1 \ldots n$, as follows:

$\Gamma = \{\delta_{ij} \mid \delta_{ij} \in \Omega_i \forall i = 1 \ldots n \forall j = 1 \ldots p_i\}$, where $\Omega_i = \delta_1 \vee \delta_2 \vee \ldots \vee \delta_{p_i}$;

execute the procedure GA(ψ) by using Γ as the initial population;

Output: concept description Ω learned by the group;

Fig. 4.3. Procedure DLS

4.6 Experimental Results

4.6.1 Experimental Design

The following parameter values for the GA were selected after fine-tuning the GA over several runs. The GA implemented in DLS uses uniform crossover operator with a probability of 0.7. The interval range for each attribute on the rule is considered as a single entity for the purpose of crossover. The mutation operator used can be thought of as a specialization/generalization operator, which either increases or decreases (with equal probability) the interval range of an attribute by either increasing or decreasing the upper or lower interval limit with equal probability of 0.1. The reproduction operator uses a tournament selection with a size of 2. The initial population is created by generating random disjuncts. A population size of 100 is used and the terminating criterion for the GA is the non-improvement in the fitness value of the best individual in 10 generations. All the algorithms used in this chapter were implemented in C++ on a Sun workstation running SunOS 5.7.

The systems were tested on a real world chemical process control plant data. The data had 30 process variables, of which only 9 were controllable, and one boolean classification variable. All variables had continuous values from the domain [0.0 0.99]. The data set had 5720 instances, of which 3550 were positive examples and 2170 were negative examples. It was randomly broken up into 10 pairs of a training set of 3440 examples and a testing set of 2280 examples. Four different problem variations were created from the above data set. In the first case only the 9 controllable variables were used. In the second case all the 30 variables were used, and for the last two cases 20 variables were constructed by randomly combining the existing variables to create a data set with 50 and 70 variables.

Having these four different sizes of essentially the same problem allows us to test the performance of the system as the size of the problem domain increases and study its scale-up properties, something very crucial for data mining applications. The last two problem variations also allow us to test the effectiveness with which the data mining system can detect the irrelevant attributes. Typical data mining applications involve a lot of irrelevant attributes and it is widely recognized that around 80% of the resources in such applications are spent on cleaning and preprocessing the data.

4.6.2 Effectiveness of Feature Selection

We carried out several experiments to first test the effectiveness of using the binary selection vector for performing feature selection. Since we are using a population-based method (GA), one of the best ways to study the effectiveness of such a technique is to study the evolution of population proportions having different features. We can conclude that the feature selection technique is effective if the proportion of individuals having features that are important or relevant keeps growing and the proportion of individuals having features that are not relevant keeps dwindling. In other words, the proportion of good features in a population should grow and that of irrelevant features should decline with generations. We can also be sure of the robustness of this method if the proportion of different features in the population converges to similar levels irrespective of the initial proportion of those features in the population.

Feature Proportions

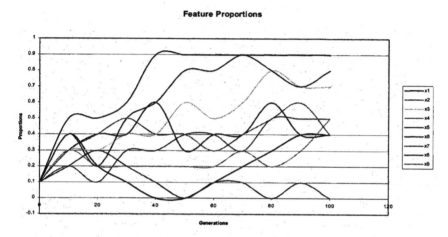

Fig. 4.4. Evolution of feature proportions with Fselect=0.1

Several runs of the GA, without feature selection, produced concepts that always included attribute X_8 with a very high discriminating interval indicating that X_8 was an extremely important attribute. To test the feature selection method outlined above we carried out experiments on the 9-attribute problem mentioned earlier. The proportion of each feature (attribute) selected in a population was tracked and three experiments were carried out with the initial proportion of each feature in the population set at 0.1, 0.5, and 0.9 (i.e., the parameter *Fselect* was set at 0.1, 0.5, 0.9 respectively). Figs. 4.4, 4.5 and 4.6 show the convergence of the proportions for the 9 features. Irrespective of the initial feature proportions it can be seen from the three figures that features X_8, X_5, and X_3 quickly take over indicating that they might be very important, and feature X_7 dies out of the population indicating that it might not be relevant. In fact, the proportion of individuals having a feature can also indicate the relative importance of that feature. For e.g., from the figures we can see that X_8 is consistently present in the maximum proportion of individuals in the population followed by X_5 and X_3. The other features tend to be present in only about half the population indicating that they are not very important and might be used only in refining the rules. Their actual importance would be reflected by the corresponding intervals in the

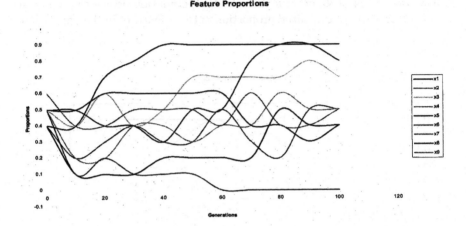

Fig. 4.5. Evolution of feature proportions with Fselect=0.5

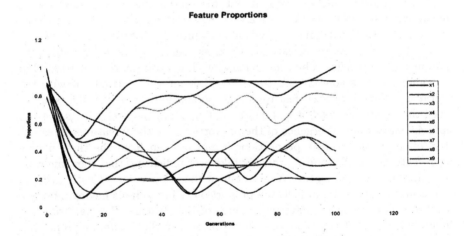

Fig. 4.6. Evolution of feature proportions with Fselect=0.9

rules that the GA is evolving simultaneously. For e.g., in one trial the best rule given by the GA was as follows:

X_1	X_2	X_3	X_4	X_5	X_6	X_7	X_8	X_9
(0.0 0.76)	(0.0 0.77)	(0.48 0.99)	(0.003 0.99)	(0.45 0.99)	(0.0 0.89)	(0.1 0.86)	(0.0 0.72)	(0.1 0.68)
0	0	1	1	1	1	0	1	0

The above rule covered 1130 positive examples and 330 negative examples out of the 2280 examples of the testing set. Note that although the selection vector selected the features X_3, X_4, X_5, X_6, and X_8, we can see that the feature X_4 is not very important for the rule as its interval range of (0.003, 0.99) almost covers its entire domain of (0.00, 0.99).

This experiment shows that the feature selection method presented above is effective in learning the important and the irrelevant features. Similar results have been observed for the above problem domain with 30, 50, and 70 features. In the next section we present detailed experimental results evaluating the performance of DLS.

4.6.3 Effectiveness of DLS

In the next set of experiments the DLS system was tested in which GAs are used as the learning agents so that the feature selection technique embedded within the GA can be used at the front end of the system. Attribute-based decomposition is used in these experiments as we want to study the scale-up properties of the system when the problem size is increased. For comparison purposes a single agent version of the system is also tested where the GA is used as the sole learning agent. To delineate the effect of using feature selection with that of using distributed learning on the performance of the system, a single agent version of the system with feature selection is also tested. The same parameter values are used for the GAs as described in section 6.1. As before 10 different pairs of data sets are used and for each data set pair the system is run 5 times with different random number seeds. The performance of the system is measured in terms of the prediction accuracy (%), computation time (sec.), and the rule size.

The detailed results for the 9 variable, 30 variable, 50 variable, and the 70 variable problems are presented below in Tables 4.1-4.4. Note that each cell is an average of 5 trials. The single agent version of the system without feature selection is used as a benchmark and the one-tailed Student-t test for paired results is performed to test the statistical significance of the results. The average and standard deviation values are provided together with the t-values of the test wherever the differences are significant at 0.05 level of significance. The average results from the four problem sizes are summarized in Table 4.5 to provide a clearer picture about the scale-up properties of the methods.

Some of the major findings can be summarized as follows:

Table 4.1. Results for 9 variable problem

Data Set	Simple GA w/o Feature Selection (GA)			GA with Simultaneous FS (GAFS)			DLS (No. of Agents = 2)		
	%	Sec.	Size	%	Sec.	Size	%	Sec.	Size
1	78.77	20.67	4	78.86	9.61	3.2	78.07	30.89	3
2	83.63	24.52	4	82.10	9.97	3.6	79.82	29.07	3
3	81.32	22.24	4.2	80.52	10.36	3.8	78.51	33.77	4
4	80.26	21.78	3.6	80.52	8.89	3.6	76.75	36.00	4
5	83.95	21.83	3.4	83.49	10.84	3.8	83.77	32.17	3
6	85.52	24.71	3.8	84.02	9.78	3.8	83.77	35.79	4
7	79.56	21.14	4	80.61	11.90	4	80.70	32.85	3
8	81.32	21.04	3.8	80.79	10.94	3.4	81.14	34.78	4
9	82.55	23.11	3.6	81.14	8.54	3.4	79.39	33.19	4
10	81.4	21.31	3.6	81.75	9.97	3.4	81.58	31.20	3
Avg.	**81.73**	**22.24**	**3.8**	**81.38**	**10.08**	**3.6**	**80.35**	**32.97**	**3.5**
Std.	*2.02*	*1.43*	*0.25*	*1.52*	*0.99*	*0.25*	*2.32*	*2.22*	*0.53*
Paired *t*-Test (2 tail)									
P_{GA}				-	1.3E-8	-	0.026	4.3E-7	-
P_{GAFS}							0.04	0.4E-9	-

Table 4.2. Results for 30 variable problem

Data Set	Simple GA w/o Feature Selection (GA)			GA with Simultaneous FS (GAFS)			DLS (No. of Agents = 3)		
	%	Sec.	Size	%	Sec.	Size	%	Sec.	Size
1	83.77	96.46	4	81.84	43.67	4	82.90	80.57	3.20
2	82.63	92.74	4	85	48.88	4.8	85.23	103.9	4.20
3	80.53	100.42	3.8	80.44	38.47	3.6	87.72	102.0	4.00
4	79.56	86.88	3.6	78.69	42.46	3.8	78.51	87.71	4.00
5	83.77	87.17	3.6	77.72	46.18	4.6	82.76	97.02	4.00
6	83.51	70.32	5	80.62	44.35	4	82.69	100.6	4.00
7	87.11	85.79	4.8	86.41	45.04	4.6	85.61	94.39	3.40
8	82.72	67.54	4.2	80.79	44.37	3.8	85.44	99.39	4.40
9	81.23	74.14	4.4	78.86	42.03	3.6	82.37	90.34	3.40
10	78.95	60.91	3.6	77.72	34.97	2.8	82.72	94.95	4.00
Avg.	**82.38**	**82.24**	**4.1**	**80.81**	**43.04**	**3.96**	**83.6**	**95.09**	**3.86**
Std.	*2.41*	*13.25*	*0.5*	*2.94*	*3.93*	*0.6*	*2.52*	*7.21*	*0.39*
Paired t-Test (2 tail)									
P_{GA}				-	3.9E-6	-	-	0.03	-
P_{GAFS}							0.01	4.8E-9	-

Table 4.3. Results for 50 variable problem

Data Set	Simple GA w/o Feature Selection (GA)			GA with Simultaneous FS (GAFS)			DLS (No. of Agents = 5)		
	%	Sec.	Size	%	Sec.	Size	%	Sec.	Size
1	79.21	114.47	4	80.09	89.67	4	83.60	94.97	4.4
2	79.21	114.13	3.8	81.41	94.22	4.4	85.96	87.40	4
3	77.02	95.39	3.2	78.59	73.31	3.4	83.86	85.33	3.8
4	77.11	88.79	3	79.04	91.49	4.4	83.07	89.72	3.8
5	78.69	102.95	3.6	79.12	76.26	3.4	82.55	83.33	3.4
6	80.44	101.75	3.8	81.84	88.60	4.4	83.25	86.19	3.6
7	80.97	105.35	3.4	83.95	91.73	4.4	87.90	92.83	3.8
8	79.30	111.71	4.2	81.06	102.23	4.8	84.65	86.13	3.8
9	78.33	108.92	3.8	78.25	77.06	3.6	83.42	92.98	4
10	77.81	108.84	4	75.26	76.33	3.4	79.47	79.27	3.2
Avg.	78.81	105.23	3.68	79.86	86.09	4.02	83.77	87.81	3.78
Std.	1.30	8.28	0.38	2.38	9.68	0.53	2.20	4.85	0.33
Paired t-Test (2 tail)									
P_{GA}				-	3.6E-4	-	1.1E-5	1.8E-4	-
P_{GAFS}							1.2E-6	-	-

Table 4.4. Results for 70 variable problem

Data Set	Simple GA w/o Feature Selection (GA)			GA with Simultaneous FS (GAFS)			DLS (No. of Agents = 7)		
	%	Sec.	Size	%	Sec.	Size	%	Sec.	Size
1	73.68	180.46	5.0	77.02	118.69	4.2	83.16	121.49	4.4
2	77.20	178.14	5.2	79.47	111.78	4.0	85.18	110.20	4
3	72.98	157.55	4.6	76.05	114.43	4.2	82.81	113.26	4.2
4	71.93	158.53	4.4	75.44	126.41	4.8	83.16	115.25	4.2
5	75.61	167.70	4.8	75.88	115.66	4.0	82.28	108.44	3.6
6	78.51	164.20	5.0	80.27	130.55	5.0	83.33	109.12	3.8
7	76.93	181.48	4.6	80.00	133.74	5.0	86.32	116.76	4.2
8	76.14	149.61	4.0	78.42	121.21	4.6	81.14	112.88	4
9	74.04	170.55	5.0	78.25	129.16	5.0	81.32	111.23	4
10	71.49	170.09	5.2	75.00	134.41	5.6	80.96	123.22	4.4
Avg.	**74.85**	**167.83**	**4.78**	**77.58**	**123.60**	**4.64**	**82.97**	**114.18**	**4.08**
Std.	*2.38*	*10.55*	*0.38*	*1.96*	*8.32*	*0.53*	*1.73*	*5.04*	*0.25*
Paired *t*-Test (2 tail)									
P_{GA}				3.2E-5	1.7E-6	-	7.8E-7	4.8E-8	8E-4
P_{GAFS}							4.9E-6	0.005	0.007

Table 4.5. Summary of results

No. of Variables	Simple GA w/o Feature Selection (GA)			GA with Simultaneous FS (GAFS)			DLS		
	%	Sec.	Size	%	Sec.	Size	%	Sec.	Size
9	81.73	22.24	3.8	81.38	10.08	3.6	80.35	32.97	3.5
30	82.38	82.24	4.1	80.81	43.04	3.96	83.6	95.09	3.86
50	78.81	105.23	3.68	79.86	86.09	4.02	83.77	87.81	3.78
70	74.85	167.83	4.78	77.58	123.60	4.64	82.97	114.18	4.08

- Performance of the simple GA quickly deteriorates as the problem size increases. Its prediction accuracy plummets at the same time its computational time and the complexity of the concept increase substantially.
- When self-adaptive feature selection is added to the simple GA (GAFS), the performance in terms of computation time and the prediction accuracy of the concept does improve significantly without significant change in the complexity of the concepts learned. However, the general performance trend as the problem size increases remains the same, with a degradation in performance as the problem size increases from 30 variables to 70 variables. As hypothesized earlier, the main advantage of the self-adaptive feature selection method is to improve the computational time required, but it is interesting that in doing so it also improves the prediction accuracy of the concepts learned. For example, for the 70 variable problem the GAFS method improves the prediction accuracy by statistically significant 4% and reduces the computational time by about 26%.
- The DLS method, in general, improves the overall performance. However, it can be seen that the benefit of distributed learning is realized and becomes more pronounced as the problem size is increased. Because of the overhead costs of data distribution, communication, and synthesis involved in distributed learning, it is not beneficial to use DLS for smaller problems. However, these costs become insignificant in comparison to the actual cost of data mining as the problem size increases. For example, for the 70 variable problem, the DLS improves the prediction accuracy by a statistically significant 10%, reduces the computational time by about 32%, and reduces the rule size by about 15% compared to the GA.

4.7 Conclusions and Future Work

In this chapter we presented a genetic algorithm based multi-agent data mining system called Distributed Learning System (DLS) and evaluated its performance on four different sizes of a real world problem. We introduced a novel technique of feature selection that can be embedded in a GA that simultaneously does rule learning. We clearly demonstrated the potential of the DLS in providing good scale-up property and improved performance as measured by the prediction accuracy of its classifiers.

Currently work is under way in extending the system on several fronts. In the problem decomposition phase of the system a mix of the attribute-based and record-based decomposition is being investigated. In the local problem-solving phase we are working on incorporating different data mining algorithms. In the synthesis phase, we are investigating the use of genetic programming (GP), instead of a simple GA, that will help in extending the representation language of the classifiers beyond the disjunctive normal form.

Also, one of the advantages of using probabilistic population based methods like GA/GP is that sampling of the training data set can be introduced in the function evaluation component improving the efficiency of the system without adversely deteriorating its performance. We are currently investigating the use of sampling. Preliminary results with the same problem domain showed that the efficiency of the DLS can be improved by as much as 13% with 75% sampling without significantly degrading the prediction accuracy of the learned classifiers.

References

4.1 Angeline, P. J. (1995): Adaptive and self-adaptive evolutionary computations, in: Palaniswami, M., Attikiouzel, Y., Marks, R., Fogel, D., Fukuda, T., Eds., Computational Intelligence: A Dynamic Systems Perspectives, 152-163. (IEEE Press. Piscataway, NJ)

4.2 Angeline, P. J. (1996): Two self-adaptive crossover operations for genetic programming, in: Angeline, P., Kinnear, K., Eds., Advances in Genetic Programming II, 152-163 (MIT Press. Cambridge, MA)

4.3 Babcock, C. (1994): Parallel processing mines retail data, Computer World, 6

4.4 Back, T. (1992): Self-adaptation in genetic algorithms, in: Varela, F. J., Bourgine, P., Eds, Towards a Practice of Autonomous Systems: Proceedings of the First European Conference on Artificial Life, 263-271. (MIT Press, Cambridge, MA)

4.5 Bala, J., De Jong, K., Huang, J., Vafaei, H., Wechsler, H. (1992): Hybrid learning using genetic algorithms and decision trees for pattern classification, Proceedings of 14th Intl. Joint Conf. on Artificial Intelligence

4.6 Bala, J., De Jong, K., Pachowicz, P. (1994): Multistrategy learning from engineering data by integrating inductive generalization and genetic algorithms, in: Michalski, R., Tecuci, G., Eds., Machine Learning: A Multistrategy Approach Volume IV (Morgan Kaufmann, San Francisco)

4.7 Battiti, R. (1994): Using mutual information for selecting features in supervised neural net learning, IEEE Transactions on Neural Networks. 5, 537-550

4.8 Bond, A., Gasser, L. (1988): Readings in distributed artificial intelligence, (Morgan Kaufmann)

4.9 Bradley, P. S., Mangasarian, O. L., Street, W. N. (1998): Feature selection in mathematical programming. INFORMS Journal on Computing. 10

4.10 Breiman, L., Friedman, J., Olshen, R., Stone, C. (1994): Classification and regression trees. (Wadsworth, Belmont, California)

4.11 Chellapilla, K., Fogel, D. B. (1997): Exploring self-adaptive methods to improve the efficiency of generating approximate solutions to travelling salesman problems using evolutionary programming, in: Angeline, P. J., Reynolds, R. G., McDonnell, J. R., Eberhart, R., Eds., Evolutionary Programming VI, (Springer)

4.12 Cover, T. M. (1974): The best two independent measurements are not the two best, IEEE Transactions on Systems, Man, and Cybernetics, 4, 116-117

4.13 Dasarathy, B. (1991): Nearest neighbor (NN) norms: NN pattern classification techniques. IEEE Computer Society Press (Los Alamitos, CA)

4.14 Dzeroski, S. (1996): Inductive logic programming and knowledge discovery in databases, Advances in Knowledge Discovery and Data Mining, 117-152, (AAAI Press, Menlo Park, CA)

4.15 Elashoff, J. D., Elashoff, R. M., Goldman, G. E. (1967): On the choice of variables in classification problems with dichotomous variables, Biometrika, 54, 668-670

4.16 Elder, J., Pregibon, D. (1996):A statistical perspective on knowledge discovery in databases, Advances in Knowledge Discovery and Data Mining, 83-113, (AAAI Press, Menlo Park, CA)

4.17 Elomaa, T., Ukkonen, E. (1994): A geometric approach to feature selection, Proceedings of the European Conference on Machine Learning, 351-354

4.18 Fogel, D. B., Fogel, L. J., Atmar, J. W. (1991): Meta-evolutionary programming, in: Chen, R. R., Ed., Proceedings of 25th Asilomar Conference on Signals, Systems, and Computers 540-545, Pacific Grove, CA

4.19 Friedman, J. (1989): Multivariate adaptive regression splines, Annals of Statistics, **19**, 1-141

4.20 Guerra-Salcedo, C., Whitley, D. (1999): Genetic approach to feature selection for ensemble creation, in Proc. of the Genetic and Evolutionary Computation Conference, 236-243

4.21 Huhns, M. N. (1987): Distributed artificial intelligence, Pitman, London

4.22 John, G., Kohavi, R., Pfleger, K. (1994): Irrelevant features and the subset selection problem, Proceedings of the 11th International Conference on Machine Learning. 121-129. Morgan Kaufmann, San Francisco

4.23 Kolodner, J. (1993): Case-based reasoning, Morgan Kaufmann, San Francisco

4.24 Kira, K., Rendell, L. A. (1992): A practical approach to feature selection, Proceedings of the 9th International Conference on Machine Learning. 249-256. Morgan Kaufmann, San Francisco

4.25 Kittler, J. (1975): Mathematical methods of feature selection in pattern recognition, International Journal of Man-Machine Studies. **7**, 609-637

4.26 Koza, J. R. (1994): Genetic programming II MIT Press, Cambridge, MA

4.27 Lanzi, P. (1997): Fast feature selection with genetic algorithms: a filter approach, in Proc. of IEEE Intl. Conf. on Evolutionary Computation, 537-540

4.28 Lee, I., Sikora, R., Shaw, M. (1995): A genetic algorithm based approach to flexible flow-line scheduling with variable lot sizes, IEEE Transactions on Systems, Man, and Cybernetics, **27B**, 36-54

4.29 Levenick, J. (1991): Inserting introns improves genetic algorithm success rate: taking a cue from biology, in: Belew, R., Booker, L., Eds., Proc. of the Fourth Intl. Conf. on Genetic Algorithms, 123-127 (Morgan Kaufmann, San Mateo, CA

4.30 Nordin, P., Francone, F., Banzhaf, W. (1996): Explicitly defined introns and destructive crossover in genetic programming, in: P. Angeline and K. Kinnear, Eds., Advances in Genetic Programming: Volume **2**, 111-134, MIT Press, Cambridge, MA

4.31 Opitz, D. (1999): An evolutionary approach to feature set selection, in Proc. of the Genetic and Evolutionary Computation Conference, 803

4.32 Pearl, J. (1988): Probabilistic reasoning in intelligent systems, Morgan Kaufmann, San Francisco

4.33 Prodomidis, A. L., Chan, P. K., Stolfo, S. J. (2000): Meta-learning in distributed data mining systems: issues and approaches, in Advances in Distributed and Parallel Knowledge Discovery, Kargupta, H., Chan, P. (Eds), AAAI Press, 81-113

4.34 Provost, F. (2000): Distributed data mining: scaling up and beyond, in Advances in Distributed and Parallel Knowledge Discovery, Kargupta, H. and Chan, P. (Eds), AAAI Press, 3-27

4.35 Provost, F., Kolluri, V. (1999): A survey of methods for scaling up inductive algorithms, Data Mining and Knowledge Discovery, **2**, 131-169

4.36 Punch, W., Goodman, E., Pei, M., Chia-Shun, L., Hovland, P., Enbody, R. (1993): Further research on feature selection and classification using genetic algorithms, in: S. Forrest, Ed., Proceedings of the 5th International Conference on Genetic Algorithms, 557-564, Morgan Kaufmann

4.37 Quinlan, J. (1992): C4.5: Programs for machine learning, Morgan Kaufmann, San Francisco

4.38 Raymer, M., Punch, W., Goodman, E., Sanschagrin, Kuhn, L. (1997): Simultaneous feature scaling and selection using a genetic algorithm, in Proc. of the 7th Intl. Conf. On Genetic Algorithms, 561-567

4.39 Schwefel, H. P. (1981): Numerical optimization of computer models, Wiley, Chichester

4.40 Sikora, R. (1992): Learning control strategies for a chemical process: a distributed approach, IEEE Expert, 35-43

4.41 Sikora R., Shaw, M. (1994): A double-layered learning approach to acquiring rules for classification: integrating genetic algorithms with similarity-based learning, ORSA Journal on Computing, 6, 174-187

4.42 Sikora, R., Piramuthu, S. (1996): An intelligent fault diagnosis system for robotic machines, International Journal of Computational Intelligence and Organizations, 1, 144-153

4.43 Smith, R. G., Davis, R. (1981): Frameworks for cooperation in distributed problem solving, IEEE Transactions on Systems, Man, and Cybernetics. SMC-11, 61-70

4.44 Toussaint, G. T. (1971): Note on optimal selection of independent binary-valued features for pattern recognition, IEEE Transactions on Information Theory, 17, 618

4.45 Turney, P., (1997): How to shift bias: lessons from the baldwin effect, Evolutionary Computation, 4, 271-295

4.46 Vafaie, H., De Jong, K. (1994): Improving a rule induction system using genetic algorithms, in: Michalski, R., Tecuci, G., Eds., Machine Learning: A Multistartegy Approach Volume IV, Morgan Kaufmann, San Francisco

4.47 Whittaker, J. (1990): Graphical models in applied multivariate statistics, Wiley, NY

4.48 Yang, J., Honavar, V. (1998): Feature subset selection using a genetic algorithm, IEEE Intelligent Systems, 13, 44-49

5. A Rule Extraction System with Class-Dependent Features

Xiuju Fu[1] and Lipo Wang[2]

[1] Institute of High Performance Computing, Science Park 2, 117528, Singapore
fuxj@ihpc.a-star.edu.sg
[2] School of Electrical and Electronic Engineering, Nanyang Technological University (NTU), Nanyang Avenue, Singapore 639798.
elpwang@ntu.edu.sg
http://www.ntu.edu.sg/home/elpwang

Abstract: In the context of rule extraction, irrelevant or redundant features in data sets often impede the rule extraction computational efficiency and detrimentally affect the accuracy of rules. Data dimensionality reduction is desirable as a preprocessing procedure for rule extraction. We propose a rule extraction system for extracting rules based on class-dependent features which is selected by genetic algorithms (GAs). The major parts of the rule extraction system are: (1) class-dependent feature selection, (2) RBF neural network classifiers with class-dependent features, (3) rule extraction. The objectives of this chapter are: (1) to identify the essential characteristics of our proposed rule extraction system (2) to study the various choices in a data preprocessing procedure and (3) to explore the concept of class-dependent feature selection.

In this chapter, first, we overview rule extraction systems from the view point of its components. Then we propose a decompositional rule extraction method based on RBF neural networks. In the proposed rule extraction method, rules are extracted from trained RBF neural networks with class-dependent features. GA is used to determine the feature subsets corresponding to different classes. Rules are extracted from trained RBF neural networks by a gradient descent method.

5.1 Overview

5.1.1 Rule Extraction Systems

Huge amounts of data have been stored in documents or in hard disks of computers. Data mining [5.33, 5.37] is very useful in economic and scientific domains. Knowledge discovery from databases (KDD) techniques are used to reveal critical information hidden in data sets. As one of important tasks in KDD, rule extraction has attracted much attention in recent years.

The goal of a rule extraction system is to obtain insights for numerical data sets, and further describe and visualize the concepts of data. Generally, a rule extraction system shown in Fig. 5.1 includes the follow components:

1. *Data collection*

   Data are collected in various domains, such as in aerospace, banking and finance, retail and marketing, etc. Valuable information is hidden in huge volumes of data, which calls for intelligent and efficient techniques for discovering knowledge in order to make better decisions, improve profits, save resources, and reduce labor costs.

2. *Data preprocessing*

   Diverse data formats and data objects are stored in data repositories. Many variables (attributes) are collected for the purpose to illustrate the concept of objects. However, not all attributes are necessary for analyzing data, i.e., some irrelevant or unimportant data may be included into data sets. In order to remove irrelevant information which may interfere the data analysis process, data dimensionality reduction (DDR) is widely explored for both memory constraint and speed limitation. Hence, the follow preprocessing for data is needed.

   – *Feature selection*: Much research work [5.23, 5.25, 5.26] has been carried out in choosing a feature subset to represent the concept of data with the removal of irrelevant and redundant features.

   – *Normalization*: Text inputs have to transformed into numerical ones. For neural networks, input values are usually normalized between [0,1].

3. *The selection of rule extraction tools*

   Decision trees, neural networks, and genetic algorithms, etc., are often used as tools for rule extraction.

   – *Neural networks*: Since neural networks are excellent at predicting, learning from experiences, and generalizing from previous examples, many researchers focus on applying neural networks in the area of rule extraction [5.17, 5.27, 5.43, 5.45]. However, a disadvantage of neural networks is that it is difficult to determine neural network architectures and train parameters. Explaining the operation of a trained network is also difficult.

   – *Decision trees*: Decision trees [5.12, 5.49] can form concise rules in contrast to neural networks. However, the accuracy of decision trees is often lower than neural networks for noisy data and it is difficult for decision trees to tackle dynamic data. Decision trees can work together with neural networks for the rule extraction task. Zhao [5.51] constructed a decision tree with each node being an expert neural network for obtaining the advantages of both the decision tree and the neural network. Tsang et al [5.48] and Umano et al [5.49] combined neural networks with decision trees to obtain better performance in rule extraction.

   – *GA*: Due to its ability to search globally for the optimal solution to a problem, GA has often been combined with neural networks in rule extraction tasks. Fukumi and Akamatsu [5.9] used GA to prune the connections in neural networks before extracting rules. Hruschka and

Ebecken [5.14] proposed clustering genetic algorithm (CGA) to cluster the activation values of the hidden units of a trained neural network. Rules were then extracted based on the results from CGA. Ishibuchi et al [5.15]-[5.18] used GA to obtain concise rules by selecting important members from the rules obtained from a neural network.

4. *Expression of the extracted rules*

Usually, the rules are in IF-THEN forms. The premise parts of rules are composed of different combinations of inputs. There are 3 kinds of rule decision boundaries:

- *hyper-rectangular*
- *hyper-plane*
- *hyper-ellipse*

They are shown in Fig. 5.2-5.4. The hyper-rectangular boundary is the simplest. However, since the distributions of data may be different for different problems, different decision boundaries or combinations of different boundaries may be required (see Fig. 5.5) for different problems. Finding the most efficient decision boundary type will be one of future tasks.

By rule extraction techniques, people can learn what neural networks have generalized from data sets and how neural network models predict and estimate, which is useful in breaking the black-box curse of neural networks. Neural networks can be applied more widely in diverse rule extraction techniques.

5.1.2 Categories of Rule Extraction Systems

Many methods have been proposed for rule extraction. These rule extraction systems can be characterized by:

1. Form of allowed inputs: continuous, discrete, or both continuous and discrete variables.
2. Form of extracted rule decision boundaries: hyper-rectangular, hyper-plane, hyper-ellipse
3. Approach for searching rules: pedagogical and decompositional approaches [5.42]

There are quite a few of methods dealing with discrete variables [5.42, 5.46] or continuous variables [5.45, 5.47]. Only a few of methods deal with both continuous and discrete variables [5.7, 5.3]. Some rule extraction methods extract rules with hyper-plane decision boundaries [5.13, 5.9], and some with hyper-rectangular rule decision boundaries [5.3, 5.16, 5.27]. Rules with hyper-ellipse decision boundaries can be obtained from RBF-based rule extraction methods directly, however, the complexity of extracted rules makes it unpopular.

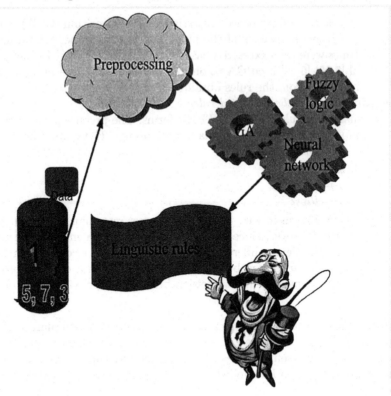

Fig. 5.1. A rule extraction system

The pedagogical algorithms consider a neural network to be a black box and use only the activation value of input and output units in the neural network when extracting rules through a neural network. In contrast, the decompositional algorithms consider each unit in a neural network and unify them into the rules corresponding to the neural network. Compared with the former algorithms, the later ones can utilize each hidden unit of neural networks, and can obtain detail rules [5.44].

5.1.3 Data Dimensionality Reduction

In many application areas, knowledge is discovered from large databases using data mining methods. With static rule extraction methods, in most cases, it is evident that the accuracy of extracted rules is better and the size of data can be reduced to save the computation burden if fewer features are selected. And DDR procedure may reduce the number of features that need to be collected.

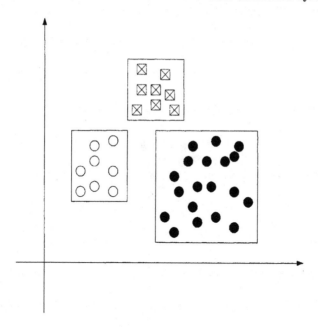

Fig. 5.2. Hyper-rectangular decision boundary

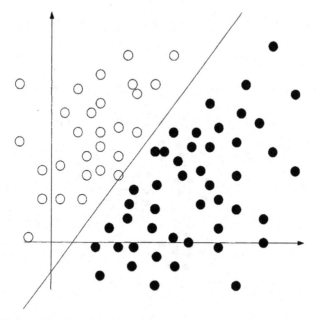

Fig. 5.3. Hyper-plane decision boundary

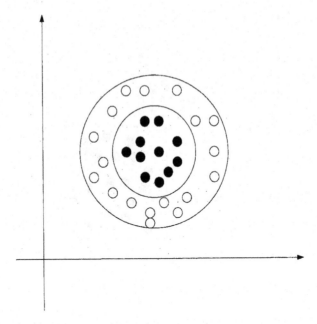

Fig. 5.4. Hyper-ellipse decision boundary

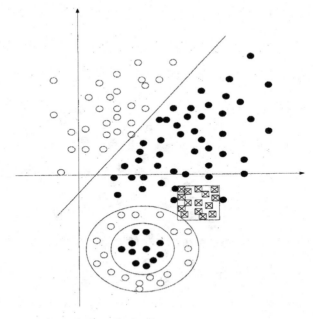

Fig. 5.5. Decision boundaries of mixed types

Feature extraction and feature selection. DDR is to map high-dimensional patterns onto lower-dimensional patterns. Techniques for DDR may be classified into two categories: feature extraction and feature selection.

Feature extraction creates a number of new features through a transformation of the raw features. Linear Discriminant Analysis (LDA) [5.23, 5.25, 5.26] and Principal Components Analysis (PCA) [5.21] are two popular techniques for feature extraction. Though the transformations are designed to maintain concepts in the data, it is difficult to prevent by-products from affecting detrimentally the original concepts in the data.

Feature selection techniques select the best subset of features out of the original set. Feature selection is desirable since it does not generate new features or unwanted by-products. The attributes which are important to maintain the concepts in the original data are selected from the entire attribute set. How to determine the importance level of attributes is the key to feature selection techniques. Mutual-information-based feature selection (MIFS) [5.25, 5.1] is a common method of feature selection, in which "the information content" of each attribute (feature) is evaluated with regard to class labels and other attributes. By calculating mutual information, the importance level of features are ranked based on their ability to maximizing the evaluation formula. However, in MIFS, the number of features to be selected need to be pre-defined.

Filter and wrapper approaches. In feature selection algorithms, there are two basic categories. The first is the filter approach [5.5, 5.50] which sieves a suitable feature subset based on a fitness criterion, such as the inconsistency between the feature subset with class labels.

The second is the wrapper approach [5.19, 5.31, 5.32, 5.38]. In the wrapper approach, feature selection is wrapped in the induction algorithm. The feature subset is selected during the reasoning process of an induction algorithm. In [5.32], the importance factor of each input feature of a multi-layer perceptron (MLP) neural network is determined by the weighted connections between the input and the second layer of the MLP during training. The features with importance factors below a certain level are eliminated.

The difference of the two algorithms lies in whether or not the feature selection is carried out independently of induction algorithms. Sometimes, the filter approach can not efficiently remove the irrelevant features because it totally ignores the effect of the selected feature subset on the performance of induction algorithms. The wrapper approach can be time consuming especially for those induction algorithms that are computationally intensive, such as neural networks. [5.50] combines the filter and the wrapper approaches to reduce time complexity and improve classification accuracy.

Class-independent and class-dependent feature selection. With the consideration that a feature has its own expertise for discriminating different classes, feature selection may be classified into class-dependent feature selection and class-independent feature selection. In data mining applications, a clear explanation about what causes a certain behavior or a development trend is the first concern of clients. In clustering or classification tasks, people would like to know what leads to the results more than the prediction results. Which attribute play a key role is unknown in those tasks.

In past work, it is assumed that all classes in a data set share the same features, hence, DDR tasks usually focus on finding general features for all classes. The individual discriminatory capability of each feature is ignored. However, the individual discriminatory capability of each feature is very important information hidden in the data. Take a simple example, when an insurance company evaluate which customer is with high credit confidence, the personal information of the customer may include too many items. Finding which item affects the customer credit most may help the insurance company in making decision and track the customer information mainly on the item, which may reduce company cost, and improve corresponding response speed according to the information collected.

Class-independent feature selection techniques are widely developed in research area. A feature subset is selected according to the evaluation score. The evaluation score is the average score of the features included in the subset for discriminating classes from each other.

GA is often combined with other methods to select class-independent features [5.4, 5.5, 5.10, 5.24, 5.39], popular searching algorithms. Each chromosome in the population pool represents a feature mask [5.4, 5.5, 5.24, 5.39]. The number of binary bits in a GA string equals to the number of features. If the binary bit is 1 at a certain position represent that the corresponding feature is selected, else the feature is excluded in the feature set selected.

Oh et al [5.35, 5.36] proposed class-independent feature selection to improve recognition performance. Class-dependent features were selected by considering class separation in conjunction with the recognition rate [5.36]. Next, Oh et al constructed multiple MLP classifiers based on the class-dependent features obtained. For each class, a MLP classifier whose inputs were the features selected for this class was trained individually. Thus, if there are M classes in the data set, M MLP classifiers have to be trained, which is computationally expensive.

In [5.20], features are selected according to their ability in classifying each pair of classes. The individual capability of each feature is taken into consideration for discriminating different pairs of classes. The nature of class-dependent features is detected. However, finally, a general group of features is chosen in which the features with high discriminatory abilities are all included, and the individual abilities of features are hidden. In a multi-class data set, each class may have its own feature subset which distinguishes this class well from other classes and represents the characteristics of this class. And with the presence of redundant or irrelevant features for each class in data sets, it is desirable to select class-dependent features corresponding to each class. In this work, we propose to select class-dependent features using GA based on a novel radial basis function (RBF) classifier. Since each feature may have different capabilities in discriminating different classes, features are masked differently for different classes. In our RBF classifier, each Gaussian kernel function of the RBF neural network is strongly active for only a subset of patterns which are approximately of the same class. A group of Gaussian kernel functions is generated for each class. In our method, different feature masks are used for different groups of Gaussian kernel functions corresponding to different classes. The feature masks are adjusted by GA. The classification accuracy of the RBF neural network is used as the fitness function. Thus, the feature subsets which better distinguish each class from other classes are obtained and the dimensionality of a data set is reduced.

5.2 Our Rule Extraction System

In this section, linguistic rules are extracted from our novel RBF neural network classifier with class-dependent features. For different groups of hidden units corresponding to different classes in RBF neural networks, different feature subsets are selected as inputs. GA is used to search for the optimal feature masks for all classes. In contrast to Oh *et al* [5.35, 5.36], only a single such RBF network, rather than multiple MLPs, are required for a multi-class problem when selecting class-dependent features.

The rule extraction system based on class-dependent feature selection includes:

1. *Data collection:* Data from www.ics.uci.edu are used for testing our algorithm.
2. *Data preprocessing:* Each attribute of data is normalized between [0,1]. Select class-dependent features using GA from RBF neural networks.
3. *The selection of rule extraction tools:* RBF neural network classifiers are constructed based on class-dependent features using GA. The information of data is embedded in the trained RBF neural network architecture. The gradient descent method is used for extracting rules based on the RBF neural network architecture.
4. *The expression of the extracted rules:* The rules extracted will be in IF-THEN forms. Compact rules with few premises and hyper-rectangular rule boundaries are generated.

In order to state the rule extraction system clearly, a conventional RBF classifier and our novel RBF classifier are described and compared in next section. In this section class-dependent feature selection technique will be presented together with the construction of the novel RBF classifier. The rule extraction procedure is described in Section 5.4.

5.3 RBF Classifiers

5.3.1 A Conventional RBF Classifier

In a conventional RBF neural network [5.2, 5.11, 5.22, 5.40, 5.41], all features are used as inputs for function approximation, pattern classification, prediction, etc.. The default weights from the input layer to the hidden layer are 1's. If there are M classes in the data set, the m-th output of the network is represented as follows:

$$y_m(\mathbf{x}) = \sum_{j=1}^{K} w_{mj} \emptyset_j(\mathbf{x}) + w_{m0} b_m. \tag{5.1}$$

Here $\mathbf{x}$ is the n-dimensional input pattern vector, $m = 1, 2, ..., M$, K is the number of hidden units. M is the number of output. w_{mj} is the weight connecting the j-th hidden unit to the m-th output node. b_m is the bias. w_{m0} is the weight connecting the bias and the m-th output node. $\emptyset_j(\mathbf{x})$ is the activation function of the j-th hidden unit:

$$\emptyset_j(\mathbf{x}) = e^{\frac{-||\mathbf{x} - \mathbf{C_j}||^2}{2\sigma_j^2}}, \tag{5.2}$$

where $\mathbf{C_j}$ and σ_j are the center and the width for the j-th hidden unit, respectively, which are adjusted during learning.

The Euclidean distance between the input vector and the center vector is used to measure the activation response of a hidden unit to the input pattern. A hidden unit represents a cluster of patterns and is active remarkably only when the input pattern falls within this cluster. Most patterns in the subset are with the same class label as the initial center of the hidden unit. Thus, the hidden unit mainly serves the class. In this RBF neural network training algorithm, the initial centers are randomly selected from the training set. The weights connecting hidden units and output units will show the relationship between hidden units and classes. Among the weights connecting a hidden unit to all output units, the weight with the largest positive magnitude shows which output unit the hidden unit mainly serves.

Usually, multiple hidden units are generated for one class. The hidden units serving class i are not generated continuously. A hidden unit for class i can be generated after generating a cluster for class j. But we can group them to show that every class corresponds to a subset of hidden units as in Fig. 5.6.

It is desirable to choose feature subsets for each class, which can lead to better separation of classes by removing interference caused by the presence of redundant and irrelevant features. A novel RBF classifier is thus constructed, which we will describe in next section.

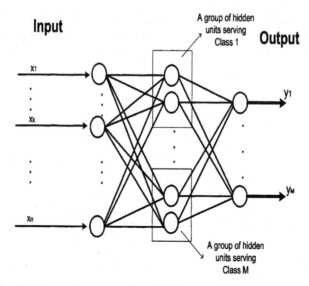

Fig. 5.6. Architecture of a conventional RBF neural network

5.3.2 A Novel RBF Classifier

In this section, we propose a novel RBF classifier with class-dependent features and describe its training algorithm.

We observe that the hidden neurons in an RBF network may be grouped according to classes. That is, if most of the patterns in the cluster represented by a hidden neuron belong to class i, we say that this hidden neuron belongs to the group for class i (Fig. 5.7). We add a class-dependent feature mask for each group of hidden neurons. $\{x_1^i, x_2^i, ..., x_{k_i}^i\}$ (k_i is the number of class-dependent features for class i) are the features selected for discriminating class i from other classes.

The m-th output of the network is as follows:

$$y_m(\mathbf{x}) = \sum_{i=1}^{M} \sum_{j=1}^{k_i} w_{mj}^i \phi_j^i \mathbf{x} + w_{m0} b_m, \qquad (5.3)$$

where M is the number of classes, k_i is the number of hidden units serving ith class. w_{mj}^i is the weight connecting the jth hidden unit of the mth class to the ith output unit. b_m is the bias. w_{m0} is the weight connecting the bias and the mth output node. $\phi_j^i(\mathbf{X})$ is the activation function of the jth hidden unit which serves class i:

$$\phi_j^i(\mathbf{x}) = e^{\frac{-||\mathbf{x}^i - \mathbf{c}_j^i||^2}{2\sigma_j^{i\,2}}}. \qquad (5.4)$$

Here $\mathbf{x^i} = \{g_1^i x_1, g_2^i x_2, ..., g_k^i x_k, ..., g_n^i x_n\}$. $\{g_1^i, g_2^i, ..., g_k^i, ..., g_n^i\}$ is the feature mask for class i. $g_k^i = 0, 1$. σ_j^i is the width for the j-th hidden unit of class i and is obtained during training in the presence of the feature masks. $\mathbf{C_j^i} = \{g_1^i c_1, g_2^i c_2, ..., g_k^i c_k, ..., g_n^i c_n\}$.

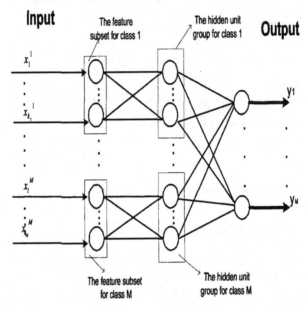

Fig. 5.7. Architecture of a new RBF neural network with class-dependent feature masks

The training algorithm for a novel RBF classifier with class-dependent features is reported in our paper [5.8]. Besides class-dependent features selected in the novel RBF classifier, another advantage of the classifier is that if a new class is added into the data set, there is no need to train the entire RBF classifier. All the hidden units in the old classifier can be maintained. The training procedure can begin from Step 2 above. After the hidden units for the new class are obtained, the new hidden units can be combined with the previous hidden units and the weights between the hidden layer and the output layer can be calculated using the LLS method. Similarly, if some new patterns of an existed class are added to the data set, there is no need to train the entire RBF classifier either. All the hidden units corresponding to other classes in the old classifier can be maintained. The training procedure can begin from Step 2 above to search for clusters of the class concerned. After the hidden units for the class are obtained, the new hidden units can be combined with those maintained hidden units. The weights between the hidden layer and the output layer can then be calculated using the LLS method.

5.3.3 Feature Masks Encoded by GA

Due to its powerful capability in solving optimization problems, GA has been widely used in many applications. GA can be used to search for optimal solutions based on its fitness evaluation and offspring generation strategies. Usually, a binary string (an individual in the population pool) is used to represent a solution for a problem. Each individual is evaluated by the defined fitness function. The operators of GA, such as selection, crossover and mutation, are used for producing offsprings. Parents with higher fitness score are given high probabilities to generate offsprings.

The applications of GA in class-dependent feature selection can be found in the literature [5.4, 5.5, 5.10, 5.24, 5.39]. In our novel RBF classifier training algorithm, we use GA for determining class-dependent feature subsets. Suppose n is the total number of the original features and M is the number of classes. A binary string representing a possible solution in GA is shown in Fig. 5.8. The length of each individual is $n \cdot M$ bits. A chromosome G is presented as follows:

$$G = \{(g_1^1, ...g_i^1, ..., g_n^1), ..., (g_1^k, ...g_i^k, ..., g_n^k), ..., (g_1^M, ...g_i^M, ..., g_n^M)\}. \quad (5.5)$$

Here $g_i^k = 0, 1$. $k = 1, 2, ..., M$. $i = 1, 2, ..., n$.

GA operators: crossover and mutation. In the procedure for selecting class-dependent feature subsets, the roulette wheel selection is used to select chromosomes in each generation. In the roulette wheel selection, the selection probability is proportional to each chromosome's fitness. Two-point crossover is used. Two points are randomly located in each of the two parents. The two parts of the parent chromosomes between the two pairs of points are then exchanged to generate new offsprings. The probability of crossover is usually around 80%.

In the evolutionary computation by GA, mutation operator is used to prevent the fixation at some particular loci of the parent chromosomes. A locus in the parent chromosome is selected randomly and the bit at the position is replaced, i.e., if the original bit is 0, it is replaced by 1, and vice versa. Usually, the mutation rate is relatively small to avoid too much variation. However, at later generations, the number of identical members increases, which leads to a stagnant state. In order to break stagnant states to search for optimal results, we use a dynamic mutation rate, i.e., if the number of identical members in a population exceeds a certain percentage, the mutation rate is increased by a certain amount.

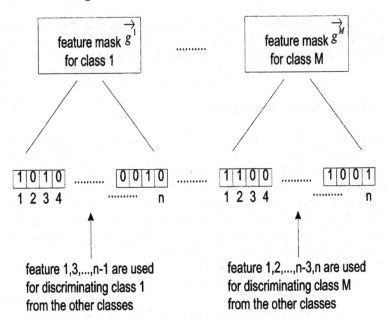

Fig. 5.8. An encoding string presenting an exemplar solution

Fitness function. Our fitness function is:

$$F(G) = 1 - E_v(G) \qquad (5.6)$$

where $E_v(G)$ is the classification error rate of the validation data set for chromosome G.

5.4 Rule Extraction by the Gradient Descent Method

The objective of tuning the rule premises is to determine the boundaries of rules so that a high rule accuracy is obtained for the test data set. In this section, we propose to extract rules from trained RBF neural networks using the gradient descent method.

Before starting the tuning process, all of the premises of the rules must be initialized. Assume that the number of attributes is n. The number of rules equals to the number of hidden neurons in the trained RBF network. The number of the premises of rules equals to n. The upper limit U_{ji} and the lower limit L_{ji} of the jth premise in the ith rule are initialized according to the trained RBF classifier as:

$$U_{ji}^{(0)} = \mu_{ji} + \sigma_i, \qquad (5.7)$$

$$L_{ji}^{(0)} = \mu_{ji} - \sigma_i, \qquad (5.8)$$

where μ_{ji} is the jth item of the center of the ith kernel function. σ_i is the width of the ith kernel function.

We introduce the following notations. Suppose $\eta^{(t)}$ is the tuning rate at time t. Initially $\eta^{(0)} = 1/N_I$, where N_I is the number of iteration steps for adjusting a premise. N_I is set to be 20 in our experiments, i.e., the smallest changing scale in one tuning step is 0.05, which is determined empirically. E is the rule error rate.

$$Q_{ji}^{(t)} \equiv \frac{\partial E}{\partial U_{ji}}|_t, \tag{5.9}$$

$$A_{ji}^{(t)} \equiv \frac{\partial E}{\partial L_{ji}}|_t . \tag{5.10}$$

$U_{ji}^{(t)}$ and $L_{ji}^{(t)}$, the upper and lower limits at time t, are tuned as follows.

$$U_{ji}^{(t+1)} = U_{ji}^{(t)} + \Delta U_{ji}^{(t)}, \tag{5.11}$$

$$L_{ji}^{(t+1)} = L_{ji}^{(t)} + \Delta L_{ji}^{(t)}. \tag{5.12}$$

Initially, we let

$$\Delta U_{ji}^{(0)} = \eta^{(0)}. \tag{5.13}$$

$$\Delta L_{ji}^{(0)} = -\eta^{(0)}. \tag{5.14}$$

Subsequent $\Delta U_{ji}^{(t)}$ and $\Delta L_{ji}^{(t)}$ are calculated as follows.

$$\Delta W_{ji}^{(t)} = \begin{cases} \eta^{(t)} & , \text{ if } Q_{ji}^{(t-1)} < 0 \\ -\eta^{(t)} & , \text{ if } Q_{ji}^{(t-1)} > 0 \\ \Delta W_{ji}^{(t-1)} & , \text{ if } Q_{ji}^{(t-1)} = 0 \\ -\Delta W_{ji}^{(t-1)} & , \text{ if } Q_{ji}^{(t-1)} = 0 \text{ for} \\ & , \quad \frac{1}{3}N_I \text{ consecutive} \\ & , \text{ iterations,} \end{cases} \tag{5.15}$$

where $W = U, L$. When $Q_{ji}^{(t)} = 0$ consecutively for $\frac{1}{3}N_I$ time steps, this means that the current direction of premise adjustment is fruitless. $\Delta W_{ji}^{(t)}$ changes its sign as shown in the 4th line of Eq. (5.15). In this situation, we also let $\eta^{(t)} = 1.1\eta^{(t-1)}$, which helps to keep the progress from being trapped. Otherwise $\eta^{(t)}$ remains unchanged.

Compared with the technique proposed by McGarry [5.28, 5.29, 5.30], a higher accuracy with concise rules is obtained in our method. In [5.28, 5.30], the input intervals in rules are expressed in the following equations:

$$X_{upper} = \mu_i + \sigma_i - S, \tag{5.16}$$

$$X_{lower} = \mu_i - \sigma_i + S. \tag{5.17}$$

Here X_{upper} is the upper limit of the premise of one rule, and X_{lower} is the lower limit. S is feature "steepness", which was discovered empirically to be about 0.6 by McGarry. μ_i is n-dimensional center location of rule i, and σ_i is the width of receptive field. We note that the empirical parameter S may not be suitable to all data sets.

Two rule tuning stages are used in our method. In the first tuning stage, the premises of m rules (m is the number of hidden neurons of the trained RBF network) are adjusted using gradient descent theory for minimizing the rule error rate. Some rules do not make contribution to the improvement of the rule accuracy, which is due to what stated following. The input data space is separated into several subspaces through training the RBF neural network. Each subspace is represented by a hidden neuron of the RBF neural network and is a hyper-ellipse. The decision boundary of our rules is hyper-rectangular. We use gradient descent for searching the premise parts of rules. Since overlaps (Fig. 5.9(a)) exist between clusters of the same class, some hidden neurons may be overlapped completely when a hyper-rectangular rule is formed using gradient descent (see Fig. 5.9(b)). Thus, the rules overlapped completely are redundant for representing data and should be removed from the rule set. It is expected that this action will not reduce the rule accuracy. The number of rules will be less than the number of hidden neurons.

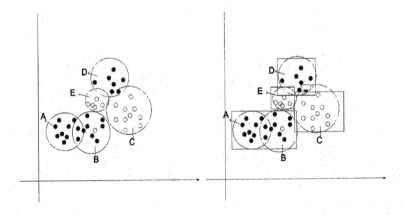

Fig. 5.9. (a) Clusters in an RBF network (b) Hyper-rectangular rule decision boundaries corresponding to the clusters

Before implementing the second tuning stage, the rules overlapped completely by other rules will be removed according to the result of the first tuning stage. At the second stage, the steps tuning premises of each rule using gradient descent are repeated again on the left rules. The rule accuracy is improved with the removing of rules overlapped and the fine tuning.

5.5 Experimental Results

Thyroid data set and Wine data set from the UCI Repository of Machine Learning Databases [5.34] are used in this section to test our algorithm

5.5.1 Thyroid Data Set

Table 5.1. Feature mask for Thyroid data set

Classes	Feature masks
Class 1	0 1 1 1 1
Class 2	0 1 1 0 0
Class 3	0 1 0 0 0

Table 5.2. Rule accuracy for Thyroid data set based on class-dependent features

rule accuracy	full features	class-dependent features
training set	94.57%	95.54%
validation set	95.35%	94.6%
testing set	90.7%	95.48%

It is shown in the feature masks (Table 5.1) that feature 1 does not play important role in discriminating classes. Hence, the T3-resin uptake test can be unnecessary in this type of Thyroid diagnosis. For class 3, feature 2 can discriminate class 3 from other classes. Feature 2 and 3 are used to classify class 2 from other classes. Feature 2, 3, 4, 5 is used to discriminate class 1 from other classes.

Two rules are extracted for Thyroid data set based on class-dependent features. The rule accuracy (in Table 5.2) is: 95.54% for training data set, 94.6% for validation data set, and 95.48% for test data set. With full features as inputs, 2 rules are obtained, and the rule accuracy is 94.57% for the test data set, 95.35% for the training data set, and 90.7% for the validation set. Thus, higher rule accuracy and more concise rules are obtained when using class-dependent features.

Rules for Thyroid data set based on class-dependent features:
Rule 1:

    IF   attribute 2 is within the interval: (12.9, 25.3)

    AND   attribute 3 is within the interval: (1.50, 10.0)

    THEN the class label is hyper-thyroid.

Rule 2:

    IF   attribute 2 is within the interval: (0.00, 5.67)

THEN   the class label is hypo-thyroid.
Default rule:
the class label is normal.

5.5.2 Wine Data Set

Table 5.3. Feature mask for Wine data set

Classes	Feature masks
Class 1	0 1 0 1 1 1 1 0 1 0 1 1 0
Class 2	0 0 1 1 1 1 1 0 0 1 1 1 1
Class 3	0 0 1 1 0 0 0 0 0 0 1 1 1

Table 5.4. Rule accuracy for Wine data set based on class-dependent features

rule accuracy	full features	class-dependent features
training set	90.6%	88.7%
validation set	77.8%	83.4%
test set	86.1%	86.1%

It is shown in the feature masks (Table 5.3) that the feature subset $\{2,4,5,6,7,9,11,12\}$ plays important role in discriminating class 1 from other classes, the feature subset $\{3,4,5,6,7,10,11,12,13\}$ is used to discriminate class 2 from other classes, and the feature subset $\{2,3,11,12,13\}$ is used to discriminate class 3 from other classes.

7 rules are extracted for Wine data set based on class-dependent features. The rule accuracy (in Table 5.4) is: 88.7% for the training data set, 83.4% for validation data set, and 86.1% for the test data set. With full features as inputs, 7 rules are obtained, and the rule accuracy is 90.6% for the training data set, 77.8% for the validation set, and 86.1% for the test set. Thus, the same rule accuracy and more concise rules with few premises are obtained when using class-dependent features compared to using full features.

5.5.3 Summary

In this chapter, we overview rule extraction systems and their preprocessing component: data dimensionality reduction, and look into our proposed rule extraction system with class-dependent feature selection. We also proposed a rule extraction method for extracting rules from a novel RBF classifier based on class-dependent features. The discriminatory power of each feature for discriminating classes is considered for each class. Different feature subsets are selected for different classes individually based on its ability in discriminating the class with other classes, which shows the relationships between the feature subset and the class concerned. The class-dependent feature selection results obtained above provide a new direction for analyzing the relationships between features and classes. The reduction in dimensionality can lead to compact rules in the rule extraction task. Thyroid and Wine data sets are used to test the algorithm. Experimental results show that our proposed algorithm is effective in reducing the number of feature input and leads to compact and accurate rules simultaneously.

References

5.1 Battiti, R. (1994): Using mutual information for selecting features in supervised neural net learning. IEEE Trans. NN, **54**, 537-550

5.2 Bishop, C.M. (1995): Neural network for pattern recognition. Oxford University Press Inc. New York.

5.3 Bologna, G., Pellegrini, C. (1998): Constraining the MLP power of expression to facilitate symbolic rule extraction. IEEE Proc. Computational Intelligence, **1**, 146-151

5.4 Brill, F. Z., Brown, D. E., Martin, W. N. (1992): Fast generic selection of features for neural network classifiers. IEEE Trans. NN, **3**, 324-328

5.5 Chaikla, N., Qi, Y. L. (1999): Genetic algorithms in feature selection. IEEE Proc. SMC, **5**, 538-540

5.6 Devijver, P. A., Kittler, J. (1982): Pattern recognition: a statistical approach. Prentice-Hall International, Inc. London

5.7 Fu, X. J., Wang, L. P. (2001): Rule extraction by genetic algorithms based on a simplified RBF neural network. IEEE Proc. CEC, **2**, 753-758

5.8 Fu, X. J., Wang, L. P. (2002): Rule extraction from an RBF classifier based on class-dependent features, IEEE Proc. CEC. **2**, 1916-1921

5.9 Fukumi, M., Akamatsu, N. (1998): Rule extraction from neural networks trained using evolutionary algorithms with deterministic mutation. IEEE Proc. Computational Intelligence, **1**, 686-689

5.10 Fung, G. S. K., Liu, J. N. K, Chan, K. H., Lau, R. W. H. (1997): Fuzzy genetic algorithm approach to feature selection problem. IEEE Proc. Fuzzy Systems, **1**, 441-446

5.11 Gomm, J. B., Yu, D. L. (2000): Selecting radial basis function network centers with recursive orthogonal least squares training. IEEE Trans. NN, **11**,, 306-314

5.12 Gupta, A., Sang, A., Lam, S. M. (1999): Generalized analytic rule extraction for feedforward neural networks. IEEE Trans. Knowledge and Data Engineering, **11**, 985-991

5.13 Hruschka, E. R., Ebecken, N. F. F. (1999): Rule extraction from neural networks: modified RX algorithm. Proc. NN, **4**, 2504-2508

5.14 Hruschka, E. R., Ebecken, N. F. F. (2000): Applying a clustering genetic algorithm for extracting rules from a supervised neural network. IEEE Proc. IJCNN, **3**, 407-412

5.15 Ishibuchi, H., Murata, T., Turksen, I. B. (1995): Selecting linguistic classification rules by two-objective genetic algorithms. IEEE Proc. SMC, **2**, 1410-1415

5.16 Ishibuchi, H., Nii, M. (1996): Generating fuzzy if-then rules from trained neural networks: linguistic analysis of neural networks. IEEE Proc. ICNN, **2**, 1133-1138

5.17 Ishibuchi, H., Nii, M., Murata, T. (1997): Linguistic rule extraction from neural networks and genetic-algorithm-based rule selection. Proc. ICNN, **4**, 2390-2395

5.18 Ishibuchi, H., Murata, T. (1998): Multi-objective genetic local search for minimizing the number of fuzzy rules for pattern classification problems. IEEE Proc. Computational Intelligence, **2**, 1100-1105

5.19 Jain, A., Zongker, D. (1997): Feature selection: evaluation, application, and small sample performance. IEEE Trans. PAMI, **19**, 153 -158

5.20 Ji, H., Bang, S. Y. (2000): Feature selection for multi-class classification using pairwise class discriminatory measure and covering concept. Electronics Letters, **36**, 524 -525

5.21 Kambhatla, N., Leen, T. K. (1993): Fast non-linear dimension reduction. IEEE Proc. NN. **3**, 1213 -1218

5.22 Kaminski, W., Strumillo, P. (1997): Kernel orthonormalization in radial basis function neural networks. IEEE Trans. NN, **8**, 1177 - 1183

5.23 Kawatani, T., Shimizu, H. (1998): Handwritten kanji recognition with the LDA method. Proc. Pattern Recognition, **2**, 1301 -1305

5.24 Kuncheva, L. I., Jain, L. C. (2000): Designing classifier fusion systems by genetic algorithms. IEEE Trans. Evolutionary Computation, **4**, 327 -336

5.25 Kurt, D. B., Ghosh, J. (1996): Mutual information feature extractors for neural classifiers. IEEE Proc. IJCNN, **3**, 1528 -1533

5.26 Liu, C. J., Wechsler, H. (1998): Enhanced fisher linear discriminant models for face recognition. Proc. Pattern Recognition, **2**, 1368 -1372

5.27 Lu, H. J., Setiono, R., Liu, H. (1996): Effective data mining using neural networks. IEEE Trans. Knowledge and Data Engineering, **8**, 957 -961

5.28 McGarry, K. J., Wermter, S., MacIntyre, J. (1999): Knowledge extraction from radial basis function networks and multilayer perceptrons. Proc. IJCNN, **4**, 2494-2497

5.29 McGarry, K. J., Tait, J., Wermter, S., MacIntyre, J. (1999): Rule-extraction from radial basis function networks. Proc. Artificial Neural Networks, **2**, 613-618

5.30 McGarry, K. J., MacIntyre, J. (1999): Knowledge extraction and insertion from radial basis function networks. IEE Colloquium on Applied Statistical Pattern Recognition (Ref. No. 1999/063), 15/1-15/6

5.31 Mao, J. C., Mohiuddin, K., Jain, A. K. (1994): Parsimonious network design and feature selection through node pruning. Proc. Pattern Recognition, **2**, 622-624

5.32 Matecki, U., Sperschneider, V. (1997): Automated feature selection for MLP networks in SAR image classification. Proc. Image Processing and Its Applications, **2**, 676 - 679

5.33 Matheus, C. J., Chan, P. K., Piatetsky-Shapiro, G. (1993): Systems for knowledge discovery in databases. IEEE Trans. Knowledge and Data Engineering, **5**, 903-913

5.34 Murphy, P. M., Aha, D. W. (1994). UCI repository of machine learning databases. Irvine, CA: University of California, Department of Information and Computer Science.

5.35 Oh, I. S., Lee, J. S., Suen, C. Y. (1998): Using class separation for feature analysis and combination of class-dependent features. Proc. Pattern Recognition, **1**, 453 - 455

5.36 Oh, I. S., Lee, J. S., Suen, C.Y. (1999): Analysis of class separation and com-bination of class-dependent features for handwriting recognition. IEEE Trans. PAMI, **21**, 1089-1094

5.37 Pazzani, M. J. (2000): Knowledge discovery from data?. IEEE Intelligent Sys-tems, **15**, 10-12

5.38 Pudil, P., Ferri, F. J., Novovicova, J., Kittler, J. (1994): Floating search meth-ods for feature selection with nonmonotonic criterion functions. 12th IAPR Proc. Computer Vision and Image Processing, **2**, 279 -283

5.39 Raymer, M. L., Punch, W. F., Goodman, E. D., Kuhn, L. A., Jain, A. K. (2000): Dimensionality reduction using genetic algorithms. IEEE Trans. Evo-lutionary Computation, **4**, 164-171

5.40 Roy, A., Govil, S., Miranda, R. (1995): An algorithm to generate radial basis function (RBF)-like nets for classification problems. Neural networks, **8**, 179-201

5.41 Roy, A., Govil, S., Miranda, R. (1997): A neural-network learning theory and a polynomial time RBF algorithm. IEEE Trans. NN, **8**, 1301-1313

5.42 Saito, K., Nakano, R. (1998): Medical diagnostic expert system based on pdp model. IEEE Proc. NN, **1**, 255-262

5.43 Saito, T., Takefuji, Y. (1999): Logical rule extraction from data by maximum neural networks Intelligent Processing and Manufacturing of Materials. Proc. Intelligent Processing and Manufacturing of Materials, **2**, 723-728

5.44 Sato, M., Tsukimoto, H. (2001): Rule extraction from neural networks via decision tree induction. Proc. IJCNN, **3**, 1870 -1875

5.45 Setiono, R. (1997) :Extracting rules from neural networks by pruning and hidden-unit splitting. Neural Computation, **9**, 205-225

5.46 Setiono, R., Leow, L. W. (2000): FERNN: An algorithm for fast extraction of rules from neural networks. Applied Science,**12**, 15-25

5.47 Thrun, S. (1995): Extracting rules from artificial neural networks with dis-tributed representations. Advances in Neural Information Processing Systems, **7**, MIT Press, Cambridge, MA

5.48 Tsang, E. C. C., Wang, X. Z., Yeung, D. S. (1999): Improving learning accu-racy of fuzzy decision trees by hybrid neural networks. IEEE Proc. SMC, **3**, 337-342

5.49 Umano, M., Okada, T., Hatono, I., Tamura, H. (2000): Extraction of quantified fuzzy rules from numerical data. IEEE Proc. Fuzzy Systems, **2**, 1062-1067 .

5.50 Yuan, H., Tseng, S. S., Wu, G. S., Zhang, F. Y. (1999): A two-phase feature selection method using both filter and wrapper. IEEE Proc. SMC, **2**, 132-136

5.51 Zhao, Q.F. (2001): Evolutionary design of neural network tree-integration of decision tree, neural network and GA. Proc. Evolutionary Computation, **1**, 240 -244

6. Knowledge Discovery in Data Mining via an Evolutionary Algorithm

Qi Yu, Kay Chen Tan, and Tong Heng Lee

Department of Electrical and Computer Engineering National University of Singapore 4 Engineering Drive 3, Singapore 117576.

Abstract: One of the major challenges in data mining is the extraction of comprehensible knowledge from recorded data. In this chapter, a coevolution-based classification technique, namely CORE (COevolutionary Rule Extractor), is proposed to discover cohesive classification rules in data mining. Unlike existing approaches where candidate rules and rule sets are evolved at different stages in the classification process, the proposed CORE coevolves rules and rule sets concurrently in two cooperative populations to confine the search space and to produce good rule sets that are cohesive and comprehensive. The proposed coevolutionary classification technique is extensively validated upon three data sets obtained from UCI machine learning repository, which are representative artificial and real-world data from various domains. Comparison of results show that the proposed CORE produces comprehensive and good classification rules for most data sets, which are competitive as compared to existing classifiers in literature. Simulation results obtained from box plots and t-tests also unveil that CORE is relatively robust and invariant to random partition of data sets.

6.1 Introduction

Evolutionary algorithms for knowledge discovery in data mining can be broadly divided into Michigan and Pittsburgh approaches [6.15] depending on how rules are encoded in the population of individuals. In the Michigan approach, each individual encodes a single prediction rule. Examples of EAs for classification that follow the Michigan approach are REGAL [6.8], GGP [6.30] and [6.3]. In the Pittsburgh approach, each individual encodes a set of prediction rules [6.7]. Examples of EAs for classification that follow the Pittsburgh approach are GABIL [6.4], GIL [6.10] and BGP [6.25]. The choice between the two coding approaches strongly depends on which kind of knowledge is targeted. For the task of classification, the quality of rule set is usually evaluated as a whole rather than the quality of a single rule, i.e., the interaction among the rules is important. In this case, the Pittsburgh approach is a better choice. On the other hand, the Michigan approach is more suitable for tasks where the goal is to find a small set of high-quality prediction rules, and each rule is evaluated independently of other rules [6.19].

The Pittsburgh approach directly takes into account rule interaction when computing the fitness function of an individual. However, the individuals encoded with this approach are syntactically longer, thus making the optimal solution difficult to be found when the search space is large. In Pittsburgh approach, standard genetic operators may need to be modified for coping with the relatively complex individuals to ensure feasibility of solutions. On the other hand, in Michigan approach encoded individuals are simpler and syntactically shorter, thus making the search of solutions easier and faster. However, this approach has some drawbacks, e.g., each rule is encoded and evaluated separately without taking into account interactions among different rules. Furthermore, the Michigan approach often needs to include niching methods such as token competition [6.30] in order to maintain the diversity of population or to converge to a set of rules instead of a single rule.

To utilize advantages of both approaches and to compromise the drawbacks of each, the two approaches can be applied together in a certain way. Ishibuchi et al. [6.9], proposed a multi-criteria genetic algorithm for extraction of linguistic fuzzy rules that considers both the accuracy and length of a rule set. In their approach, a pre-screening technique was employed to generate the candidate rules that are encoded with Michigan approach. These candidate rules are selected solely based upon the length of fuzzy rules without considering the applicability or usefulness of the rules. These candidate rules are then used to construct rule sets that are encoded with Pittsburgh approach. Although the number of candidate rules is reduced (587 out of 1296 possible fuzzy rules for Iris data set), the total number of candidate rules generated by the algorithm still remains large. Furthermore, since the rules and rule sets are searched at different stages, it does not necessarily guarantee the cohesiveness of the rules obtained.

One approach to ensure cohesiveness of the solutions is to evolve both elements simultaneously through coevolutionary-based algorithms, which could evolve multiple populations concurrently in data classification [6.14, 6.20]. It has been shown that by coevolving a population of fuzzy membership function with a population of GA individuals [6.20] or GP tree individuals [6.14], better results could be produced as compared to those without the coevolution. Unlike existing approaches, a coevolutionary-based rule extraction and classification system, namely CORE (COevolutionary Rule Extractor) is proposed in this chapter to coevolve different types of species, e.g., individuals of rules and rule sets in the evolutionary process. It is shown that instead of evolving random rules, the efficiency and performance of the classifier can be improved by coevolving the populations of rules and rule sets.

The coevolutionary rule extractor is empowered with token competition [6.30] to generate the pool of candidate rules. With this technique, the number of candidate rules is significantly reduced and the applicability and usefulness of the candidate rules are automatically assured by the niching capability of the token competition. The population of rules is coevolved cooperatively in parallel with a group of co-populations nurturing the rule sets. Because of the difference in targeted solutions, the Michigan and Pittsburgh coding approach is employed in the main population and co-populations respectively. The performance of the proposed CORE is extensively evaluated upon three selected data sets from UCI Machine Learning Repository (http://www.ics.uci.edu/~mlearn/MLRepository.html), which is widely used benchmark and real-world data collection in data mining and knowledge discovery community. The classification results of the proposed CORE are analyzed both qualitatively and statistically, and are compared with many widely used traditional and evolutionary based classifiers.

The rest of the chapter is organized as follows: Besides giving a general overview of coevolutionary algorithms, Section 6.2 presents the proposed coevolutionary rule extraction (CORE) algorithm. Various features in CORE such as chromosome structure, co-populations, fitness evaluation and token competition are also described in the section. The problem sets used for validation are presented in Section 6.3, and the simulation results are analyzed and discussed. Conclusions are drawn in Section 6.4.

6.2 Coevolutionary Rule Extractor

6.2.1 Coevolutionary Algorithms

Coevolution refers to the simultaneous evolution of two or more species with coupled fitness [6.13]. Such coupled evolution favors the discovery of complex solutions [6.21]. Coevolution of species can either compete (e.g., to obtain exclusivity on a limited resource) or cooperate (e.g., to gain access to some hard-to-attain resource). In a competitive coevolutionary algorithm, the fitness of an individual is based on direct competition with individuals of other species, which in turn evolve separately in their own populations. Increased fitness of one of the species implies a diminution in the fitness of the other species. This evolutionary pressure tends to produce new strategies in the populations involved to maintain their chances of survival. This "arms race" ideally increases the capabilities of all species until they reach an optimum. For further details on competitive coevolution, readers may refer to [6.24].

Cooperative coevolutionary algorithms involve a number of independently evolving species, which together form a complex structure for solving difficult problems. The fitness of an individual depends on its ability to collaborate with individuals from other species. In this way, the evolutionary pressure stemming from the difficulty of the problem favors the development of cooperative strategies and individuals. Single population evolutionary algorithms often perform poorly, manifesting stagnation, convergence to local optima and computational costliness, when they are confronted with problems presenting one or more of the following features: (1) the sought-after solution is complex, (2) the problem or its solution is clearly decomposable, (3) the genome encodes different types of values, (4) strong interdependencies among the components of the solution, and (5) components-ordering drastically affects fitness [6.20]. Cooperative coevolution addresses these issues effectively, and consequently widening the range of applications in evolutionary computing.

Paredis [6.21] applied cooperative coevolution to problems that involved finding simultaneously the values of a solution and their adequate order. In his approach, a population of solutions coevolves alongside a population of permutations performed on the genotypes of the solutions. Moriarty [6.18] used a cooperative coevolutionary approach to evolve neural networks. Each individual in one species corresponds to a single hidden neuron of a neural network and its connections with the input and output layers. This population coevolves alongside a second one whose individuals encode sets of hidden neurons (i.e., individuals from the first population) forming a neural network. The similar idea of coevolving a species of fundamental elements and a species of complex elements build from the fundamental elements are applied in the proposed CORE. In CORE, the species of fundamental elements is the population of rules and the species of the complex elements is the population of rule sets. These coevolving populations are coupled cooperatively on fitness as the quality of rule sets greatly depends on the quality of rules forming the rule sets. Details of the proposed CORE are presented in the following subsections.

6.2.2 Overall Flowchart of CORE

The overall flowchart of CORE is presented in Fig. 6.1 to give a complete overview of the coevolutionary algorithm. As can be seen, the learning process consists of two groups of populations that evolve rules and rule sets respectively and cooperatively. The evolution process of these populations is described below.

The algorithm first builds from the training data set a gene map to maintain a mapping of genes to the corresponding attributes in the data set. The main population will then be initialized according to the gene map to ensure only valid chromosomes are created. Chromosomes in the main population are encoded using Michigan approach where each chromosome represents a single rule. These chromosomes are variable in length, and all the initial chromosomes are evaluated against the training data set for their fitness before starting the iteration looping. The mating pool is first formed by selecting parents from the main population using tournament selection. The genetic operators such as crossover and mutation are then applied upon the mating pool to reproduce the offspring. The offspring are assigned as the new main population and passed into the token competition that works as a covering algorithm. The token competition effectively maintains a pool of good rules, i.e., rules that coveres the solution space well. As classification problems generally contain not only one but many useful knowledge or regulations, it is crucial for the coevolutionary algorithm to maintain a population with high

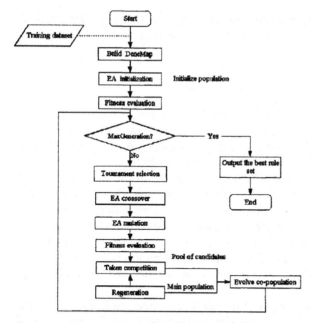

Fig. 6.1. The overall flowchart of the coevolutionary rule extractor

diversity. To achieve this, a regenerate operator is used, which replaces chromosomes that are below average fitness in the main population with randomly generated chromosomes at some user specified probability. After the regeneration, all chromosomes in the pool resulting from the token competition and one-tenth of randomly selected chromosomes from the main population will be used to create the co-populations.

The number of co-populations is determined by the maximum number of rules allowed in a rule set. For example, if a rule set is allowed to have up to 15 rules, then there will be 15 co-populations. Each co-population maintains a number of rule sets with the same number of rules. All chromosomes in the co-populations are encoded with Pittsburgh approach where each chromosome represents a rule set. The quality of these rule sets is greatly affected by the rules used. To evaluate the co-chromosomes, the classification accuracy on the training set is used. Here, only mutation operator is applied to evolve the co-chromosomes in order to avoid the reproduction of redundant rule sets. After the new main population and co-populations have been evolved, the coevolution will proceed to the next generation and the process will be repeated until the last generation is reached. At the end of the evolution, each sub-population outputs its 'best' candidate rule set, which will compete (based on the classification accuracy) with the 'best' rule sets generated by other co-populations to obtain the final optimal rule set. To retain concise rule sets in the classification, a shorter rule set is preferable to a longer one even if both achieved the same classification accuracy. In this way, the order as well as the number of rules in the rule sets can be optimized and determined simultaneously.

6.2.3 Population and Chromosome Structure

The coevolution strategy is used on two different population structures, e.g., the main population coevolves with the co-populations. The main population contains subpopulation of each class. For example, if there are 3 classes, the main population consists of 3 subpopulations, where each subpopulation evolves individual rules for each class. Chromosomes in the main population are encoded with Michigan approach, that is, one chromosome represents a rule. In a given chromosome, each gene is associated with an attribute of the data set. The number of genes used to construct the chromosome is variable. Therefore each chromosome does not need to contain all the attributes or to contain the attributes in order. Since most data sets are consisted of nominal and numeric attributes, two types of genes are included to handle them respectively.

Nominal attributes take on a finite number of discrete values while numeric attributes are those that are able to take on a continuous range of numerical values. As they are different in nature, different handling techniques are required. To do that, the gene structure is facilitated with 4 fields: attribute index, type, relation and value. The attribute index is the index of the attribute that the gene corresponds to. Type indicates whether the attribute is nominal or numeric. The relation field is used to assign the relationship operator for the gene. For nominal gene, only equal ($==$) and not equal ($\neq$) operators are used. On the other hand, for numeric gene, 6 comparison operators: greater than ($>$), greater than or equal ($\geq$), less than or equal ($\leq$), less than ($<$), in bound ($><$) and out bound ($<>$) are used. The in bound and out bound operators enable the chromosome to encode numeric attribute with range, i.e., rules with a $<$ attribute $<$ b (in bound) and attribute $<$ a and attribute $>$ b (out bound) are possible. The last field is the value of the gene that is directly related to the corresponding attribute. For nominal gene, the value is a bit string array that is associated with the index of value for the attribute. For example, if an attribute has three possible values, temperature = {low, medium, high}, then the value of [1 0 1] corresponding to the first attribute value (low) and the third attribute value (high). The value of numeric gene on the other hand is scaled real value ranging from 0 to 1. Thus, if a numeric attribute has the exact value ranged from 50 to 68, the 0 and 1 will represent 50 and 68 respectively. In the case of bound relation, the value is an array containing the lower and upper bound of the gene.

The CORE algorithm applied a group of co-populations to evolve rule sets with different number of rules. The chromosomes in these co-populations, namely co-chromosomes are encoded with Pittsburgh approach where each co-chromosome is encoded with a rule set. The basic element that builds up the co-chromosomes is the chromosome representing rules from the main population. The number of rules in a co-chromosome depends on which co-population the co-chromosome is attached to. For example, the fourth co-population will only contain co-chromosomes with 4 rules. Note that the default class is also encoded in the co-chromosome although it is not counted as a rule. This provides greater flexibility on constructing the rule sets. Since both the main and co-populations are differed from each other, the genetic operators applied are also differed as discussed in the next sub-section.

6.2.4 Genetic Operators

The genetic operators applied to the main population are crossover and mutation. Here, the genetic operations take place in two levels, i.e., chromosome level and gene level. At chromosome level, one point crossover is used for which a random crossover point is selected for each parent chromosome and genes are exchanged. The length of the offspring does not necessarily be the same as their parents'. The chromosome level mutation removes or inserts a newly generated gene onto the chromosome. At the gene level, crossover and mutation are different for nominal and numeric attributes. For nominal gene, the gene level crossover is a one-point crossover to the bit string array of the parents and the gene level mutation is a one-bit mutation to the bit string array. On the other hand, the gene level crossover for numeric gene is a standard real-coded crossover with the following equations,

$$offsrping_1 = \alpha \; parent_1 + (1 - \alpha)parent_2 \tag{6.1}$$

$$offsrping_2 = \alpha \; parent_2 + (1 - \alpha)parent_1 \tag{6.2}$$

where α is a random number. The gene level mutation is performed by replacing the parent's value with a randomly generated value, and the mutation operator ensures that the value is always valid. Only mutation operator is used in evolving the co-population, e.g., one rule is removed from or inserted to the co-chromosome. When the number of rules in a co-chromosome is changed, the co-chromosome will be moved to the correct co-population. For example, if a 5 rules co-chromosome in co-population 5 was mutated and one rule was removed from it, then it will be moved to co-population 4.

6.2.5 Fitness Evaluations

When a rule or individual is used to classify a given training instance, one of the four possible concepts can be observed: true positive (tp), false positive (fp), true negative (tn) and false negative (fn). The true positive and true negative are correct classifications, while false positive and false negative are incorrect classifications.

Using these concepts, the fitness function used in evaluating the main population of CORE is defined as,

$$fitness = penalty \times \frac{tp}{tp + fn} \times \{1 + \frac{tn}{tn + fp}\} \tag{6.3}$$

$$\text{with } penalty = \frac{N}{N + fp}$$

where N is the total number of instances in the training set; tp, fp, tn, and fn is true positive, false positive, true negative and false negative, respectively. The value of the fitness function is in the range of 0 to 2. The fitness value is 2 (the fittest) when all instances are correctly classified by the rule, i.e., when fp and fn are 0. The penalty factor is included in the fitness function to evaluate the quality of the combined individuals in the rule set. This is because Boolean sequential rule list (where rules are considered one after another) is too sensitive and has the tendency of having a large number of false positives (fp) due to the virtual OR connection among the rules. When a rule with large fp is considered first in a rule list, many of the instances will be classified incorrectly. Therefore, the fitness function should be penalized based on the value of fp, e.g., a penalty factor that tends to minimize fp is included in Eq. (6.3). Since the number of rules is fixed for each co-population, there is no need to explicitly formulate the comprehensibility in the fitness evaluation of co-populations. Indeed, the fitness function of the co-populations can be simply formulated as the classification accuracy.

6.2.6 Token Competition

Token competition is applied in CORE to evolve different multiple rules for prediction of each class in the data set as well as to preserve the diversity in the evolution [6.30]. It takes every class in the data set and seeks to find rules that cover all instances under the class if possible, and at the same time to exclude instances not in the class. Often no single rule can cover all instances of a class, and hence there is a need to discover more rules that can predict all instances of a class, but at the same time do not overlap with instances of another class. The CORE also applied the principle of controlling the quality of rules via the concept of minimum support as proposed by Tan [6.26]. Every instance in a data set is called a token, for which all chromosomes in the population will compete to capture. A chromosome has the chance to capture a token if all its antecedents match that in the token and the class in the token is the class predicted by the chromosome, i.e., a tp concept case. If more than one chromosome is eligible to capture the same token, then only the fittest chromosome will be assigned the token. The adjusted fitness is calculated for each chromosome after the token competition as given by,

$$adjusted\_fitness = fitness \times \frac{actualNumberOfCapturedTokens}{numberOfClassOccurences} \tag{6.4}$$

The term *numberOfClassOccurences* refers to the total number of instances or tokens in the data set containing the class to be predicted, while *actualNumberofCapturedTokens* refers to the total number of tokens in the data set that has been captured by the chromosome. Obviously, the relation is always true.

6.3 Case Study

6.3.1 Experimental Setup

The proposed CORE is validated based on three data sets. Each of these data sets is partitioned into two sets: a training set and a testing set (also called validation set). The training set is used to train CORE, through which its learning capability can be justified. However, a classifier that learns well does not necessary guarantee it is also good in generalization. In order to evaluate the generalization capability, the rule sets obtained by CORE are applied to testing set after the training. In order to ensure the replicability and clarity of the validation results, all experiments have been designed carefully in this study. In the total of 100 evolutionary runs on each of the three data sets, a random seed [1] , which is the same as the number of runs (i.e. the 50th run uses random seed 50), is first used to randomize the orders of data in the data sets. The randomized data is then partitioned with the first 66% as the training data and the remaining 34% as the test data.

Table 6.1 lists the parameter settings in CORE that are applied to all the three data sets. These parameters have been chosen after some preliminary experiments and then applied upon all the experiments. Therefore the settings should not be regarded as an optimal set of parameter values but rather a generalized one with which the CORE can perform well over a wide range of data sets. The CORE was programmed using the Java Developers Kit (JDK 1.3.1) from Sun Microsystems on an Intel Pentium IV 1.4 GHz computer with 128 MB SDRAM.

6.3.2 Simulation Results and Comparisons

The Fisher's Iris data set. Although the Fisher's Iris data set is a rather simple domain, it can be regarded as a multi-class (more than two classes) problem since it contains three classes. Through the performance study on this data set, the classification capability of CORE for multi-class problems could be assessed.

---

[1] The random number generator used in the experiments is provided with Sun's JDK 1.3.1 and the data set randomizer used is provided with WEKA [6.29]. Different partitioning of data sets might have resulted under different programming environments

Table 6.1. Parameter settings used in the experiments

Parameter	value
Population size	100
Co-population size	50
Number of generations	100
Number of co-populations	15
Probability of crossover	0.9
Probability of mutation	0.3
Probability of regeneration	0.5

(A) Experimental results

Table 6.2 summarizes the results of CORE on the Iris data, which include the results obtained by the default parameter settings in Table 6.1 as well as by another set of parameter settings with smaller training time (with a population and generation size of 50). One hundred independent simulation runs have been conducted for both of the settings, and the P-value is computed for testing the null hypothesis that the means of two groups of paired observations on the accuracy are equal. From the P-values in Table 6.2, it can be seen that larger population size and longer training time could result in better predictive accuracy, as expected.

In order to have a clearer view of the classification performance for the default parameter settings over the 100 independent evolutionary runs, the histograms plots for the results are shown in Fig. 6.2. Clearly, a normally distributed performance has been achieved by CORE for this data set. The classification rule set that has the highest predictive accuracy (i.e. the highest classification accuracy on the testing set) on the Iris data is presented in Table 6.3. Besides the fitness value, the support factor and confidence factor are provided as additional information to show the performance of each rule. The support factor measures the coverage of a rule, which is the ratio of the number of instances covered by the rule to the total number of instances. On the other hand, the confidence factor measures the accuracy of the rule. For a rule 'if X then Y' and with a training set of N instances, the support and confidence factors are given as,

$$support = \frac{number\ of\ instances\ with\ both\ X\ and\ Y}{N} \qquad (6.5)$$

$$confidence = \frac{number\ of\ instances\ with\ both\ X\ and\ Y}{number\ of\ instances\ with\ X}. \qquad (6.6)$$

It can be seen from the rule set in Table 6.3 that the classification rules are not ordered in any specific pattern since the rule sets were formed randomly during the evolution. However, the rules with higher fitness will be placed on top of the rules with lower fitness values in general. As shown in Table 6.3, all the rules in this rule set have relatively high accuracy and confidence factor. It can also be observed that the rule set was constructed of only petal length and petal width attributes, which clearly demonstrates the autonomous attribute selection ability of CORE.

Table 6.2. Summary of results on the Iris data set

	Default setting	The 2nd setting	P-value
Training			
Max	100%	100%	
Min	90.91%	90.91%	
Mean	97.64%	96.97%	0.01
StdDev	1.28%	2.17%	
Test			
Max	100%	100%	
Min	90.20%	90.91%	
Mean	96.61%	96.06%	0.01
StdDev	2.34%	2.59%	
Avg # rules	3.77	3.12	
Training time (s)	151	76	

Table 6.3. The best classification rule set of CORE for Iris data

	Rule	Fitness	Support factor	Confidence factor
1	IF petallength $<$ 1.9471, THEN class = Iris-setosa	2.0	0.3434	1.0
2	IF petallength $><$ [1.1, 2.0582], THEN class = Iris-setosa	2.0	0.3434	1.0
3	IF petalwidth $<>$ [0.1, 1.6747], THEN class = Iris-virginica	1.7031	0.303	0.9375
4	IF petallength $><$ [4.4422, 4.9938], THEN class = Iris-versicolor	1.4697	0.2424	0.8889
5	IF petallength $<>$ [1.2707, 4.4234], THEN class = Iris-virginica	1.8084	0.3333	0.8089
6	ELSE class = Iris-versicolor			

Accuracy = 100%

(B) Comparisons with other works

As stated in the Introduction, the proposed CORE is capable of generating comprehensible rule sets with good classification accuracy. For comparisons, two famous rule-based machine-learning algorithms C4.5 rules [6.22], e.g. J48

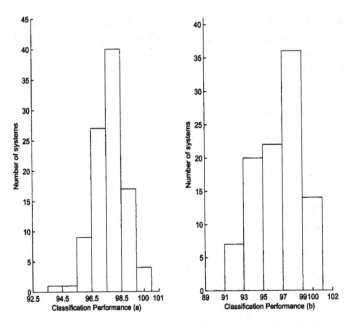

Fig. 6.2. The performance of CORE upon the Iris data (a) Training (b) Testing

in WEKA [6.29] and PART [6.6] as well as a statistical classifier Naïve Bayes [6.11] have been applied to the three data sets. The first two algorithms are chosen due to their rule-based characteristics as offered in CORE. Comparisons between these two algorithms and CORE include the performance of classification accuracy and rule set size (i.e., the number of rules in a rule set), since a good rule set should be both accurate and succinct. The method of Naïve Bayes is included here since it is a well-known statistical classifier that often gives high classification accuracy and provides good comparison to CORE in terms of classification ability. Besides the comparisons of average results and standard deviations over the 100 simulation runs, a paired t-test has also been performed between CORE and the three algorithms respectively. The P-values are computed for testing the null hypothesis that the means of the paired observations on the accuracy rate are equal.

To study the performance of CORE more thoroughly, the best and the latest results achieved by the rule-based classifiers in literature (including traditional and evolutionary approaches) according to our best knowledge are also included in the comparisons. Although such comparisons are not meant to be exhaustive, it provides a good basis to assess the reliability and robustness of CORE.

The classification results for all algorithms under comparisons are listed in Table 6.4. The co-evolutionary system (GP-Co) proposed by Mendes et al. [6.14], aimed to discover fuzzy classification rules, in which a GP evolving population and an EA evolving population are co-evolved to generate well adapted fuzzy rule sets and membership function definitions. Ten-fold cross-validation was used to test this system on the Iris data set. The GGP was proposed by Wong [6.31], which is a flexible knowledge discovery system that applied genetic programming (GP) and logic grammars to learn knowledge in various knowledge representation formalisms. A paired t-test has been carried out on the results from GGP and CORE over the 100 simulation runs. The GBML was proposed by Ishibuchi [6.9], which is a fuzzy genetic-based machine-learning algorithm that hybrids the Michigan and Pittsburgh approach. Ishibuchi [6.9] tested this algorithm on several data sets, but only the training accuracy was provided for the Iris data. The GPCE was proposed by Kishore [6.12], which is a GP-based technique dedicated to solve multi-category pattern recognition problems. In this algorithm, the n-class problem was modeled as n two-class problems and GPCE was trained to recognize samples belonging to its own class and reject samples belonging to other classes. The 50 to 50 split percent method was adopted by Kishore [6.12]as the validation scheme, and the average results on the validation set over several simulation runs are shown in Table 6.4.

As shown in Fig. 6.3, the simulation results are represented in box plot format [6.2] to visualize the distribution of simulation data in term of classification accuracy over the 100 independent runs. Each box plot represents the distribution of a sample population where a thick horizontal line within the box encodes the median, while the upper and lower ends of the box are the upper and lower quartiles. Dashed appendages illustrate the spread and shape of the distribution, and the '+' represents the outside values. It can be seen that although the best accuracy of 100% has been achieved by all algorithms, CORE is superior to other algorithms in terms of average accuracy by having the smallest performance scatter. Furthermore, CORE is shown to be statistically more stable as it produced no outlier for the 100 simulation runs.

Table 6.4. Performance comparisons for the Iris data set

Algorithm	# Rules	Average accuracy	Best accuracy	Standard deviation	P-value
CORE	3.77	96.61%	100%	2.35%	-
C4.5	3.87	93.67%	100%	3.73%	3.91×10^{-14}
PART	4.15	93.94%	100%	3.93%	1.65×10^{-14}
NaïveBayes	-	95.47%	100%	2.93%	1.40×10^{-5}
GP-Co	-	95.3%	-	7.1%	-
GGP	4	94.24%	100%	3.57%	1.88×10^{-10}
GBML	5	-	98%	-	-
GPCE	-	96%	-	-	-

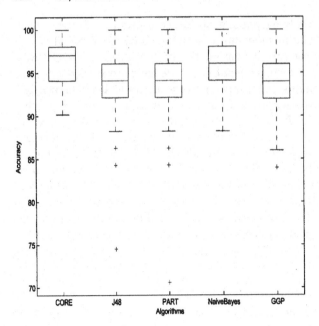

Fig. 6.3. Box plot for the Iris data set

The medical diagnosis data sets. Medical diagnosis has been known as a crucial application domain of classification in data mining. The data sets used in this study were obtained from the UCI machine learning repository. The Hepatitis data set was collected at Carnegie-Mellon University [6.1] and donated to UCI ML repository in 1988 while the Diabetes data set was firstly collected at John Hopkins University by Vincent Sigillito and then constructed by constrained selection from a large database held by the National Institute of Diabetes and Digestive and Kidney Diseases.

(A) Experimental results

Table 6.5 summarizes the classification results produced by CORE over the 100 independent runs for the medical diagnosis problems. The histograms for CORE are illustrated in Figs. 6.4 and 6.5, which again generally show a normally distributed performance. The classification rule sets that have the highest predictive accuracies on these medical data sets are presented in Tables 6.6 and 6.7. It can be observed from the relationship between the predictive accuracy of a rule set and its rule number that rule sets with larger number of rules will not necessarily lead to higher predictive accuracies. From the rule sets obtained, it is found that a typical rule set usually contains generalized rules followed by the more specific rules (which is particularly obvious on the Diabetes data set). Generally, the first several rules in a rule set will cover a great portion of the samples and left relatively few for the remaining rules. This situation is especially severe when there are many remaining rules. Therefore when the data set is not noise free, a large rule number may cause over-fitting and leads to poor generalization, which are undesirable. On the other hand, having appropriate number of specific rules could also increase the precision of the rule set as a whole, which will further enhance the classification accuracy. Having considered this situation, the designers of CORE give much flexibility to its users and let them decide whether they prefer longer rule sets or shorter ones (Users can change the number of co-populations to make the selection). If users have not much knowledge of their problem on hand, they can just discard this flexibility and use the default setting of CORE, which can still generate fairly good results.

Table 6.5. Classification results of CORE for the medical data sets

CORE	Hepatitis	Diabetes
Training		
Max	95.10%	80.24%
Min	77.45%	72.92%
Mean	88.47%	76.96%
StdDev	2.82%	1.30%
Test		
Max	92.45%	80.15%
Min	75.47%	69.85%
Mean	84.39%	75.34%
StdDev	3.72%	2.30%
Avg # rules	4.14	5.99
Training time (m)	3.55	8.18

Table 6.6. The best classification rule set of CORE for Hepatitis data

	Rule	Fitness	Support factor	Confidence factor
1	IF ASCITES = no, AND BILIRUBIN $\leq$ 1.9077, THEN Class = LIVE	1.3773	0.6373	0.9155
2	IF SPLEEN_PALPABLE $\neq$ yes, AND ASCITES $\neq$ yes, THEN Class = LIVE	1.2252	0.6275	0.8767
3	IF ALBUMIN $\leq$ 3.7234, THEN Class = DIE	1.0183	0.1569	0.4848
4	ELSE Class = LIVE			
		Accuracy = 92.45%		

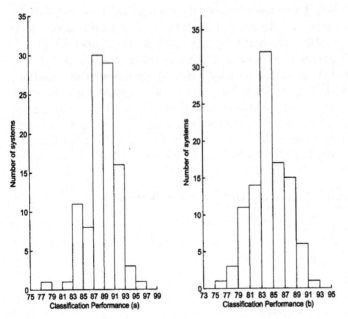

Fig. 6.4. The performance of CORE upon the Hepatitis problem (a) Training (b) Testing

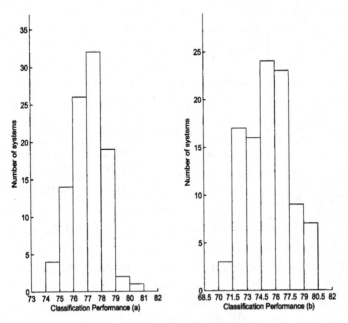

Fig. 6.5. The performance of CORE upon the Diabetes problem (a) Training (b) Testing

Table 6.7. The best classification rule set of CORE for Diabetes data

	Rule	Fitness	Support factor	Confidence factor
1	IF plas ≤ 130.7996, THEN Class = tested_negtive	1.0853	0.4881	0.7554
2	IF mass <> [5.8079, 29.9191], THEN class = tested_positive	0.8765	0.3202	0.4737
3	IF preg ≤ 4.585, AND age >< [21, 74.0987], THEN class = tested_negtive	0.8966	0.4348	0.6984
4	IF plas ≤ 138.9471, AND insu ≤ 538.5572, THEN class = tested_negtive	1.0745	0.5237	0.7341
5	IF preg>< [0, 12.2091], AND plas >< [0, 117.3103], THEN class = tested_negtive	0.9893	0.3775	0.8059
6	IF plas >< [0, 164.4243], AND age≤77.064, THEN class = tested_negtive	0.9436	0.5909	0.6765
7	IF plas > 5.8884, THEN class = tested_positive	0.6149	0.3794	0.3817
8	IF preg ≥ 4.195, AND plas < 124.8156, AND pres < 3.3973, AND skin ≤ 53.9336, AND insu >< [0, 236.2939], AND mass < 10.1695, AND pedi >< [0.078, 1.6306], AND age < 50.9739, THEN class = tested_negtive	0.0128	0.004	0.6667
9	IF preg > 7.1177, AND plas ≥ 33.5901, AND pres <> [47.0321, 78.2014], AND mass ≤ 15.1642, AND age <> [61.8127, 71.4335], THEN class = tested_positive	0.0103	0.002	1.0
10	ELSE class = tested_positive			
	Accuracy = 80.15%			

(B) Comparisons with other works

The Hepatitis data set: Wang [6.28] proposed an evolutionary rule-learning algorithm, called GA-based Fuzzy Knowledge Integration Framework (GA-based FKIF), which utilized genetic algorithms to generate an optimal or near optimal set of fuzzy rules and membership functions from the initial knowledge population. Since the average performance of GA-based FKIF was not provided in [6.28], only the best result of this algorithm is compared with CORE as given in Table 6.8. The P-values of the paired t-tests show that on this data set, CORE outperforms C4.5 rules, PART and is comparable to Naïve Bayes on the average accuracy over the 100 simulation runs, when the level of significance is set to 0.05 (α=0.05). Fig. 6.6 shows the box plot of CORE and other algorithms for comparison. As can be seen, although CORE does not achieve the best accuracy for this data set, it does produce the best average accuracy with no outlier data point.

Table 6.8. Performance comparisons for the Hepatitis problem

Algorithm	# Rules	Average accuracy	Best accuracy	Standard deviation	P-value
CORE	4.14	84.40%	92.45%	3.72%	-
C4.5	5.85	78.94%	90.57%	4.84%	4.52×10^{-22}
PART	6.64	80.02%	94.34%	4.98%	5.31×10^{-14}
NaiveBayes	-	83.62%	94.34%	4.90%	0.06
GA-based FKIF	-	-	92.9%	-	-

The Diabetes data set: The classification results for Diabetes data set from several rule-based (Itrule and CN2) and tree-based (CART, AC2 and Cal5) algorithms [6.17] are listed in Table 6.9 for comparisons. Note that these results were obtained by 12-fold cross validation. The GGP [6.31] was also applied to this data set and the results are given in Table 6.9. As can be seen from the P-values, CORE achieves a higher accuracy than GGP generally. However, as GGP is a flexible algorithm, significant performance improvement may be achieved if experts in the relevant domain can incorporate the non-trivial hidden knowledge into its predefined grammars. Fig. 6.7 shows the comparisons of CORE with other algorithms using the box plot. Clearly, CORE has achieved the best performance with very competitive classification results.

6.3.3 Discussion and Summary

The proposed CORE has been examined on three data sets obtained from UCI machine learning repository and has produced good classification results as compared to many existing classifiers. Most of the comparisons were

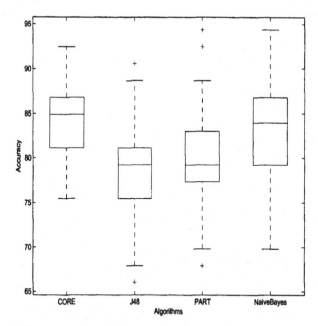

Fig. 6.6. Box plot for the Hepatitis problem

Table 6.9. Performance comparisons for the Diabetes problem

Algorithm	# Rules	Average accuracy	Best accuracy	Standard deviation	P-value
CORE	2.93	75.34%	80.15%	2.30%	-
C4.5	5.85	73.13%	77.39%	2.55%	1.83×10^{-12}
PART	6.64	72.78%	80.08%	2.53%	7.72×10^{-14}
NaiveBayes	-	75.09%	81.61%	2.54%	0.14×10^{-5}
Itrule	-	75.5%	-	-	-
CN2	-	71.1%	-	-	-
CART	-	74.5%	-	-	-
AC2	-	72.4%	-	-	-
Cal5	-	75.0%	-	-	-
GGP	14.54	72.60%	77.95%	2.97%	1.37×10^{-18}

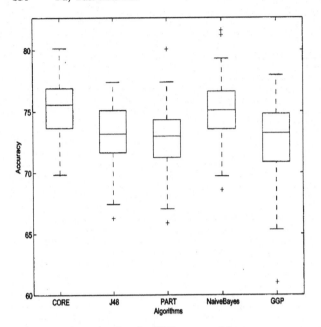

Fig. 6.7. Box plot for the Diabetes problem

performed statistically using measures such as t-tests and box plots to show the robustness of the proposed classifier. Extensive simulation results show that CORE has outperformed another two rule-based classifiers (C4.5 rule and PART) in all the testing problems and is very competitive as compared to statistical based techniques, such as Naïve Bayes. It can also be observed from the experiment results that the number of rules for the best rule set produced by CORE is relatively small as compared to other algorithms. This is an important advantage of CORE since the comprehensibility of the classification results is directly reflected by its number of rules generally. The performance comparisons to other evolutionary based classifiers (GP-Co, GGP, GBML, GPCE and GA-based FKIF) are mainly restricted by the availability of data, e.g., not all the data sets used in our experiments were tested in other publications. Since there are so many classifiers proposed in literature over the years, it is very difficult, if not impossible, to include every one of them in the comparisons. Therefore the comparisons are not meant to be exhaustive, but to assess the reliability and robustness of CORE by comparing it to some established methods widely used in the literature.

Like many other rule induction algorithms ([6.16, 6.23]), CORE employs the "separate and conquer" scheme to induction. One shortcoming of this strategy is that it causes a dwindling number of examples to be available as the induction progresses, both within each rule and for successive rules. Also, the fact that only single-attribute tests used in the rules means that all decision boundaries are parallel to the coordinate axes. With a limited number of examples available, the error of approximation to non-axis-parallel boundaries will be very large. Taking the form of decision lists, each rule body in CORE is implicitly conjoined with the negations of all those that precede it [6.5]. All of these factors will impair the performance of CORE on the problems with large number of classes. Apart from this, ordinarily, additional computation time is required by CORE in order to perform the search thoroughly, as faced by most evolutionary algorithm based approaches. Similar to many existing classifiers which are developed for off-line data classification in data mining, the proposed CORE could be very useful for many applications where the training time is less important than the accuracy and generalization abilities in the classification.

6.4 Conclusions

This chapter has proposed a cooperative coevolutionary algorithm (CORE) for rule extraction and classification in data mining applications. Unlike existing approaches, the proposed coevolutionary classifier coevolves the rules and rule sets concurrently in two cooperative populations to confine the search space and to produce good rule sets that are cohesive and comprehensive. The proposed CORE has been extensively validated upon three data sets obtained from UCI Machine Learning Repository, and the results have been analyzed both qualitatively and statistically. Comparison of results show that the proposed CORE produces comprehensive and good classification rules for all the data sets, which are very competitive or better than many classifiers widely used in literature. Simulation results obtained from the box plots and t-tests have unveiled that CORE is relatively robust and invariant to random partition of data sets. To reduce the computational effort significantly, the CORE is currently being integrated into the 'Paladin-DEC' distributed evolutionary computing framework [6.27], where multiple inter-communicating subpopulations are implemented to share and distribute the classification workload among multiple computers over the Internet.

References

6.1 Cestnik, G., Konenenko, I., Bratko, I. (1987): Assistant-86: a knowledge-elicitation tool for sophisticated users in Bratko, Lavrac (Eds.), *Machine Learning*, Sigma Press, 31-45
6.2 Chambers, J. M., Cleveland, W. S., Kleiner, B., and Turkey, P. A. (1983): Graphical methods for data analysis, Wadsworth & Brooks/Cole, Pacific CA.
6.3 De Falco, I., Della Cioppa, A., Tarantino, E. (2002): Discovering interesting classification rules with genetic programming, Applied Soft Computing, **23**, 1-13

6.4 De Jong, K. A., Spears, W. M. and Gordon, D. F. (1993): Using genetic algorithms for concept learning, Machine Learning, **13**, 161-188

6.5 Domingos, P. (1996): Unifying instance-based and rule-based induction, Machine Learning, **24**, 141-168

6.6 Frank, E. and Witten, I. H. (1998): Generating accurate rule sets without global optimization, Proceedings of the Fifteenth International Conference Machine Learning (ICML'98), 144-151

6.7 Freitas, A. A. (2003): A survey of evolutionary algorithms for data mining and knowledge discovery, to appear in: A. Ghosh and S. Tsutsui. (Eds.), Advances in Evolutionary Computation. Springer-Verlag.

6.8 Giordana, A. and Neri, F. (1995): Search-intensive concept induction, Evolutionary Computation, **3**, 375-416

6.9 Ishibuchi, H., Nakashima T. and Murata, T. (2001): Three-objective genetic-based machine learning for linguistic rule extraction, Information Sciences, **136**, 109-133

6.10 Janikow, C. Z. (1993): A knowledge-intensive genetic algorithm for supervised learning, Machine Learning, **13**, 189-228

6.11 John, G.H. and Langley, P. (1995): Estimating continuous distributions in Bayesian classifiers, Proceedings of the Eleventh Conference on Uncertainty in Artificial Intelligence, 338-345, Morgan Kaufmann, San Mateo.

6.12 Kishore, J. K., Patnaik, L. M., Mani, V. and Agrawal, V. K. (2000): Application of genetic programming for multicategory pattern classification, IEEE Transactions on Evolutionary Computation, **4**, 242-258

6.13 Liu, Y., Yao, X., Zhao, Q. F. and Higuchi, T. (2001): Scaling up fast evolutionary programming with cooperative coevolution, Proceedings of the 2001 Congress on Evolutionary Computation, IEEE Press, Piscataway, NJ, USA, 1101-1108

6.14 Mendes, R. R. F., Voznika, F. B., Freitas A. A. and Nievola, J. C. (2001): Discovering fuzzy classification rules with genetic programming and co-evolution. Principles of Data Mining and Knowledge Discovery (Proc. 5th European Conf., PKDD 2001) - Lecture Notes in Artificial Intelligence, **2168**, 314-325 Springer-Verlag

6.15 Michalewicz, Z. (1994): Genetic algorithms + data structures = evolution programs. London: Kluwer Academic Publishers.

6.16 Michalski, R. S. (1983): A theory and methodology of inductive learning. Artificial Intelligence, **20**, 111-161

6.17 Michie, D., Spiegelhalter, D. J., and Taylor, C. C. (1994): Machine learning, neural and statistical classification. London: Ellis Horwood.

6.18 Moriarty, D. E. (1997): Symbiotic evolution of neural networks in sequential decision tasks. The University of Texas at Austin.

6.19 Noda, E., Freitas, A. A., and Lopes, H. S. (1999): Discovering interesting prediction rules with a genetic algorithm. Proc. Conference on Evolutionary Computation 1999, 1322-1329, Washington D.C., USA.

6.20 Peña-Reyes, C. A., and Sipper, M. (2001): Fuzzy CoCo: A cooperative-coevolutionary approach to fuzzy modelling. IEEE Transactions on Fuzzy System, **9**

6.21 Paredis, J. (1995): Coevolutionary computation. Artificial Life, **2**, 355-375

6.22 Quinlan, J.R. (1993): C4.5: Programs for machine learning. San Mateo, CA: Morgan Kaufmann.

6.23 Rivest, R. L. (1987): Learning decision lists. Machine Learning, **2**, 229-246

6.24 Rosin, C. D. and Belew, R. K. (1997): New methods for competitive coevolution. Evolutionary Computation, **5**, 1-29

6.25 Rouwhorst, S. E., Engelbrecht, A. P. (2000): Searching the forest: using decision trees as building blocks for evolutionary search in classification databases. Proceedings of the 2000 Congress on Evolutionary Computation, 1, 633-638

6.26 Tan, K. C, Tay, A., Lee, T. H. and Heng, C. M. (2002): Mining multiple comprehensible classification rules using genetic programming. Proceedings of the IEEE Congress on Evolutionary Computation, Honolulu, Hawaii, 1302-1307

6.27 Tan, K. C., Khor, E. F., Cai, J., Heng, C. M., and Lee, T. H. (2002): Automating the drug scheduling of cancer chemotherapy via evolutionary computation. Artificial Intelligence in Medicine, 25, 169-185

6.28 Wang, C. H., Hong, T. P., Tseng, S. S. (2002): Integrating membership functions and fuzzy rule sets from multiple knowledge sources. Fuzzy Sets and Systems, 112, 141-154

6.29 Witten, I. H., and Frank, E. (1999): Data mining: practical machine learning tools and techniques with Java implementations, CA: Morgan Kaufmann Publishers.

6.30 Wong, M. L., and Leung, K. S. (2000): Data mining using grammar based genetic programming and applications. London: Kluwer Academic Publisher.

6.31 Wong, M. L. (2001): A flexible knowledge discovery system using genetic programming and logic grammars. Decision Support Systems, 31, 405-428

7. Diversity and Neuro-Ensemble

Minh Ha Nguyen, Hussein Abbass, and Robert McKay

Artificial Life and Adaptive Robotics (A.L.A.R.) Lab,
School of Information Technology and Electrical Engineering,
Australian Defence Force Academy, University of New South Wales,
Canberra, ACT 2600, Australia
m.nguyen@adfa.edu.au

Abstract : The formation of ensemble of artificial neural networks has attracted attention of researchers in the machine learning and statistical inference domains. It has been shown that combining different neural networks can improve the generalization ability of learning machines. Diversity of the ensemble's members plays a key role in minimizing the combined bias and variance of the ensemble. In this chapter, we compare between different mechanisms and methods for promoting diversity in an ensemble. In general, we found that it is important to design the diversity promoting mechanism very carefully for the ensemble's performance to be satisfactory.

7.1 Introduction

Feed-forward multilayered Artificial Neural Networks (ANNs) have been widely studied for their ability to correctly learn the true distribution of the data from a sample. This ability is called generalization. Evolutionary computation (EC), a powerful tool for global optimization, has been quite successful in training ANNs. According to the prominent review of Yao [7.50], evolutionary methods could be applied at different levels of the ANN, notably the architecture and connection weights. When EC is used for training ANNs, the process is called Evolutionary ANNs (EANNs).

Most of the work in the ANN literature concentrates on finding a single network. However, a single network found using the training set alone may not be the best network on the test set (*i.e.* it may not generalize well). The network can be either over–fitting the data or undertrained. Recently it has been found that by combining several neural networks, the generalization of the whole system could be enhanced over the separate generalization ability of the individuals [7.19, 7.25, 7.26, 7.27, 7.37, 7.41, 7.45, 7.46, 7.51, 7.52, 7.53]. The main supportive argument for the performance enhancement of the ANN ensemble is that, since different members of the ensemble possess different bias/variance trade-offs, a suitable combination of these biases/variances could result in an improvement in the generalization ability of the whole ensemble [7.41, 7.53].

This chapter presents a comparison study between a wide range of methods from literature of neuro-ensemble (*i.e.* an ensemble of neural networks).

In the rest of this chapter, we will use the words "predictor" or "classifier" interchangeable with an ANN unless it is stated otherwise.

The chapter is arranged in three sections. Section 7.2 introduces neuro-ensembles and the relationship between accuracy and diversity of the ensemble's members. Section 7.3 compares a number of methods to construct an ensemble of EANNs. The experimental results are examined across a number of dimensions. Section 7.4 summarizes our findings and presents further research questions in the field.

7.2 Ensemble of Predictors

An ensemble is a collection of predictors that can be combined to give an overall prediction. As discussed above, the problem of generalization (e.g. bias/variance dilemma) is a challenge in the machine learning field. A number of researchers showed that the overall generalization ability of a committee of experts may be better than each expert alone [7.19, 7.25, 7.26, 7.27, 7.37, 7.41, 7.51, 7.52, 7.53]. In this chapter, the experts are the ANNs. The literature of ensemble divides the assembling process into three main steps: (i) construction and training of the individual predictors, (ii) selection of suitable members to form the ensemble, and (iii) the choice of a suitable gate to combine the ensemble's members. The rest of this section is devoted to each of these three steps in neuro-ensemble.

7.2.1 Forming and Selecting the Individual Networks

A predictor is expected to give an accurate approximation of the underlying function. However, as we are learning from a sample, the best predictor on the sample may not generalize well and perform as well on unseen data. This dilemma has been contributed to the bias/variance trade-off. The best predictor on the sample would have a small bias but with a large variance. The best predictor on unseed data needs to have small bias and small variance. In order to achieve this, researchers in ensemble of predictors hypothesize that by combining the opinions of several "different" individuals or "experts", the system's generalization ability could be enhanced [7.19, 7.25, 7.26, 7.27, 7.37, 7.41, 7.45, 7.46, 7.51, 7.52, 7.53]. In the context of this chapter, the term "different" (or "diverse") refers different bias/variance trade-offs.

Consequently, the ensemble literature focuses on two important issues, namely accuracy and diversity. Accuracy refers to the ability of each individual predictor to learn and predict the correct underlying distribution of the data.

Bouckaert [7.49] defines the accuracy of an ensemble under 0-1 loss as "the one minus the 0-1 loss of the classifier, i.e. $A = 1 - L$" where the loss function is a function "that compares the prediction of a classifier with the true value of y and maps it onto a real value" (y is the data distribution). In his paper [7.49], Bouckaert derived a framework for the upper and lower bounds of the ensemble accuracy based on the mean accuracy of the individual classifiers. His proof stated that the ensemble's accuracy can never exceed twice the mean accuracy of the individuals and can never decrease below twice the mean accuracy minus 1.

Dietterich [7.14] defined an accurate classifier as the one that has an error rate of better than random guessing on new x .values. He also examined the three fundamental reasons that the ensemble works better than individuals: (i) statistically: by combining a set of classifiers, the chance of selecting a wrong classifier is reduced, (ii) computationally: a classifier with local search often gets stuck in the local minima and thus, by combining classifiers with different local search points, the system could better approximate the true unknown function, and (iii) representationally: a weighted sum of hypotheses drawn from the hypothesis space H may expand the space of representable functions.

Diversity requires that the individual members of an ensemble to be different. In many cases, accuracy and diversity are in conflict. It is often understated that selecting a set of most accurate predictors may result in an ensemble of self-similar ones. However, this is undesirable because this ensemble is not different from a single predictor. Hence, in designing neuro-ensemble, one has to take into account these two tightly coupled issues.

Researchers have been attempting to verify the relationship between diversity and the ensemble's generalization [7.23]. To solve this myth, one has to answer a number of questions: (i) how to define diversity and a diversity measure, (ii) which diversity promotion mechanisms to be used, and (iii) how can diversity be used to improve the ensemble? [7.23].

Why diversity is important?. Zenobi and Cunningham [7.54] quoted the Codorcet Jury Theorem

> If each voter has a probability p of being correct and the probability of a majority of voters being correct is M, then $p ¿ 0.5$ implies $M ¿ p$. In the limit, M approaches 1, for all $p ¿ 0.5$, as the number of voters approaches infinity.

They also stated that M is greater than p only if there is diversity in the voters and M increases with ensemble growing only if diversity is also growing. Krogh and Vedelsby [7.20] attempted to provide a framework to detach ambiguity (correlation between individuals in the ensemble) as the source of diversity to demonstrate the relationship of diversity and the networks' generalization ability ($E = \overline{E} - \overline{A}$ where $\overline{E}$ is the average errors of the individuals, $\overline{A}$ is the average disagreement among members, and E is the ensemble error). Following their argument, Rosen [7.37] and Liu et al [7.51, 7.52, 7.53, 7.27] derived a negative correlation concept to incorporate diversity to the neural network members of the ensemble. Their results showed a promising improvement in the ensemble error reduction.

However, as shown in [7.28] and in our own experiment, the diversity of the ensemble with negative correlation is very poor. McKay and Abbass [7.28] revealed that negative correlation learning tries to push the member of the ensemble away from their mean but not away from each other; therefore low diversity with negative correlation learning is not a surprise. Moreover, as noted by Kuchenva [7.23], the results of correlating different diversity measures to the ensemble errors are discouraging. There is not yet any obvious link between injecting diversity to the ensemble and the improvement of generalization ability. In other words, it is still an open question if the diversity of the ensemble does help and in which way it is useful.

Diversity measures. In the literature of ANNs and EC, there are a number of diversity measures ranging from statistical to mathematical to biologically inspired measures. Krogh and Vedelsby [7.20] defined diversity as the degree of disagreement (ambiguity) between two classifiers. Some researchers [7.13, 7.10, 7.9, 7.23] borrowed different measurements of entropy (e.g. Kullback-Leibler) in the information theory field to define diversity. Others applied statistical concepts (e.g. Kappa statistics, Q-statistics, etc..) to compute the dissimilarity between classifiers [7.23, 7.6].

Diversity could be measured on different levels of an ensemble. Kuncheva [7.23] distinguishes between the data point level and the classifier level. In the former, the diversity is computed based on the entropy of the distribution of class labels among the classifiers with respect to a certain data point, thus, the diversity of data set is the average of N within-population diversity measurements corresponding to the populations at N data points. In the latter level, some pairwise measure of diversity between two classifiers (e.g. measure of disagreement [7.23, 7.43, 7.54]) is defined and the diversity of the ensemble is the average of all possible pairs.

Diversity promotion mechanisms. The next question is how to design such a mechanism that promotes diversity in the population. There are a wide range of fields that have been studying diversity. Biologists have struggled for decades to try to understand the mechanisms causing the diversity of the biosphere (bio-diversity) existed on earth [7.44]. Evolutionary Computation researchers have been trying to solve the problem of early convergence and diversity promotion for a while. Furthermore, in the literature of ensemble of classifiers, a number of mechanisms and theoretical guidance have been proposed to promote the useful diversity among the members of an ensemble of classifiers.

7.2.2 Combining Neural Networks to form the Ensemble

Currently there are four popular combination methods, namely, majority voting, winner-take-all, simple averaging and weighted averaging. Peronne and Copper in [7.34] presented a theoretical framework for simple averaging and generalized (through weights) averaging mechanisms. Theoretically, the generalized averaging mechanism should perform better than simple averaging [7.34] in the mean square error sense. Following this intuition, Jimenez [7.19] attempt to derive a procedure to dynamically weight the certainties of the individual classifiers. Zhou et al [7.48, 7.55, 7.56] also weighted the contribution of the members into the generalization improvement, and thus, selected the highly contributing ones to form a weighted average ensemble. Fumera and Roli [7.16] theoretically and experimentally compare simple averaging and weighted averaging fusion gates. Their results showed that "weighted averaging significantly improves the performance of simple averaging only for ensembles of classifiers with highly imbalanced performance and correlation". Thus, the advantage of weighted averaging over simple averaging is quite small. This argument was supported by the experimental results in [7.46]. Pennock et al. [7.33] examined different combination gates for the classification problem. They identified several properties for the combination gates such as universality, independence of irrelevant information, scale invariance, neutrality, symmetry and positive responsiveness and their analysis showed that there are no combination gates that possess all the properties.

Beside these popular fusion methods, Sharkey [7.41] mentioned a few other methods. Sharkey divided the combination gates into four categories: (1) Averaging and Weighted Averaging, (2) Non-linear combining methods (e.g. voting and ranked-based), (3) Supra Bayesian -which is based on the probability distribution of the experts' opinions- , and (4) Stacked generalization.

7.2.3 Review of Diversity Promotion Mechanisms

Sampling data. One popular method to generate diverse members for ensemble is to sample the training data into different subsets. The purpose is to create subsets of training data such that each subset has at least some data that do not exist in other subsets. Because each subset is used to train a member network in the ensemble, it is likely that the ensemble members are different from each other, i.e. the errors are less likely related.

There are a number of techniques that can be used to sample data from a training data set. The simplest method is to divide the training data into disjoint data sets [7.41, 7.42]. However, this method works better with large data sets than with small data sets where the resultant disjointed subsets may be small enough for a neutral network to over-fit all data. Krogh and Vedelsby [7.20] used cross-validation as their sampling method. A cross-validation, also called leave-some-out, works by removing a number of data points out of the training data to form the validation set. A network is trained on the training set and tested on the validation set. A k-fold cross-validation is a mechanism in which the training set is divided into k disjoint subsets. A training is composed of $(k-1)$ subsets; and the k-th subset forms the validation set. One way to use cross-validation is to train individual networks on k different training sets and use the validation sets to estimate the generalization error. Another usage is to test the validity of a method by training the networks with k different collections of data (a collection here refers to corresponding subsets of training, validation and testing data); and then use the average error of the k collections as the true estimated error of the method.

Breiman [7.8] proposed other sampling methods called bootstrapping and bagging. It involves sampling the data set with replacement into N subsets.

Boosting and arcing. These techniques are also based on sampling the training data but the difference lies in the adaptive re-sampling scheme. In bagging, the probability given to a pattern is $\frac{1}{N}$ across N data points in the original training set, the adaptive scheme will change this probability in favor of the data that are misclassified. The original boosting method [7.38] works as follow: create the first network and train on a subset of the training data. This network is then used to select a training set for the second network by selecting a distribution of correctly classified and misclassified patterns out of the remaining patterns. Then the second trained network will jointly work with the first one to filter the patterns for the next network. In summary, this process will add a network based on the disagreement of the previous trained networks on "new" data. The original boosting method has the problem that it requires a large data set. To solve this problem, Schapire [7.15, 7.39] proposed a method called Adaboost which can be used with a small training data set by assigning a weight to each data pattern. These weights are updated in favor of the most misclassified patterns by the networks of the current round. A number of modified methods are proposed based on Adaboost. Oza [7.32] proposed an AveBoost which uses a training example weight vector that is based on the performance of all previous networks rather than the previous one alone. Kuncheva [7.21] attempted to derive a mathematical framework for the error bound of different versions of Adaboost, namely aggressive Adaboost of Conservative 1 and Conservative 2 Adaboosts. These modified versions of Adaboost are different from the original in the way of taking inward punishment action for successful and unsuccessful outcomes.

Other data varying methods. Beside the above popular methods to vary the training data and hence vary the resultant members of the ensemble, there are a number of other methods. These include distortion of input data by adding noise [7.35] or selecting different feature subsets of the feature space (Note: a data vector consists of n features and c classes - [7.54, 7.30, 7.31]). Another way is to select the input data from different data sources, (*e.g.* different sensory data [7.41, 7.42].

Tumer and Ghosh [7.45] tested four different methods based on correlation reduction: (i) the cross-validation to partition training data; (ii) pruning input features to generate new training sets; (iii) re-sampling, and (iv) weighted average of individual networks where the weights are a function of the inputs, so that in each input region a network gets more weighted than the others.

Diversity based on the correlations between the individual networks. Knogh and Vedelsby [7.20] has proposed another bias/ variance decomposition in the light of ensemble of predictors. They defined a term named "Ambiguity" which measures the disagreement of a member to the overall ensemble output (weighted average [7.20]). The ensemble ambiguity was computed as the weighted average of the individual networks' ambiguities. Thus the error of the ensemble E could be decomposed into the weighted average of the generalization error of individual networks, $\overline{E}$ minus the weighted average of the ambiguities $\overline{A}$.

$$E = \overline{E} - \overline{A}. \tag{7.1}$$

This ambiguity term captures all correlations between networks [7.20]. Their decomposition scheme suggested that increasing the negative correlation between networks could reduce the ensemble error. Based on this suggestion, a number of methods have been proposed on incorporating the negative correlation during the training of individual networks in the ensemble.

Rosen [7.37] incorporates a de-correlation measure that penalizes correlations between networks. Similarly, Liu and Yao [7.27, 7.51, 7.52, 7.53] proposed the negative correlation learning method which adds the correlation between networks as a penalty term to the error function. This penalty term is computed based on the disagreement of every network output in the ensemble with the ensemble's output (average is used). Negative correlation methods are supposed to diversify the member networks by de-correlating them. However, McKay and Abbass [7.28] showed that the networks are pushed away from their mean not from each other. Thus, it is possible that the negatively correlated networks are not actually diverse.

Diversity concepts borrowed from evolutionary computation. Another way to generate diversity in the ensemble is to borrow diversity mechanisms from other disciplines such as evolutionary computation. Lee et al [7.10, 7.9] applied speciation to diversify their evolutionary neural networks. Abbass [7.2] proposed to use the Pareto-based multi-objective method to evolve the networks that are different in respect to the objectives. He suggested a number of objectives which could be used to gain a useful diversity between networks. The first approach is to contrast the architecture's complexity against the generalization error of the ensemble. Another approach is to incorporate sampling of data where the data set is divided into subsets whose training errors serve as the objectives. The third suggestion is to inject noise to distort the output, and use the distorted output as the second objective in contrast with the original output. The second approach was found to produce the best results.

Other diversity mechanisms. As mentioned above, noise injection to the output could serve as a source of diversity. The effects of noise injection in the output level range from changing the bias/variance distribution to diversifying the models. Christensen [7.11] introduced an output distortion mechanism in which the outputs are distorted by displacing them with values from a fixed set of predetermined values. The distortion is conducted such "that the average distortion applied to each data points output values is exactly zero". Their results showed an interesting characteristic that stands out from the common belief that the individual networks in the ensemble should have low bias and high variance. He showed that large bias networks are useful too if the correlation component between them are large enough to overcome the bias component.

Most of the above ensemble methods used the same architecture for the individual networks. Another way to handle diversity issue is to vary the network's architecture. Abbass [7.2] used the number of hidden units as a possible objective in his MOP-based technique. Yao et al also introduce architecture variation in their EPNet model [7.27, 7.51, 7.52]. Renner [7.36] used cascade correlation as the constructive method; however, their selected neural networks do not require diversity on the architecture level. Thus, although using cascade correlation method, the source of diversity is actually coming from data varying techniques.

7.2.4 Review of the use of Diversity in Selecting the Members for the Ensemble

So far it is shown that diversity is widely used in constructing the individual networks for the ensemble. Another important aspect that determines the success of an ensemble technique is the pruning process (*i.e.* which members should be removed from the ensemble). Selection is an important issue because:

- using identical networks in the ensemble is meaningless. Thus, the first guide is to remove identical networks.
- networks with very low accuracy may deteriorate the whole ensemble. According to the ambiguity formula [7.20], $E = \overline{E} - \overline{A}$, including high error networks will greatly increase the first term which is undesirable unless the second term is beneficially increased more than $\overline{E}$ [7.11].
- networks that are positively correlated should not be included together. This argument is derived directly from the ambiguity formula. Positive correlation between networks means a reduction in $\overline{A}$ and thus a possible increase in the ensemble's error, E.
- few researches have shown that selection is useful for ensemble [7.55, 7.17, 7.34], or even worse, co-linearity could be very harmful to ensemble.

Diversity in Ensemble Pruning

There are different ways to make use of diversity in networks' selection. This section will briefly review some of the methods in the literature. Perrone and Cooper [7.34] introduced the use of the correlation matrix C_{ij}, where the entries represent the correlations between networks. They discussed how this correlation matrix could be ill-conditioned due to duplicate and nearly duplicate networks. Thus, it is essential to remove these duplicates. The correlation matrix, representing diversity, could be used to remove the duplicates. They also came up with an inequality condition for adding sets to ensemble.

Similarly, Zhou et al [7.55] derived a condition for removing networks from ensemble. They proposed a method called GASEN [7.55, 7.48, 7.56] that assigned weights to individuals according to the contribution of the networks to the ensemble. These weights are used to prune the networks and were evolved using GA to find an optimum contribution distribution of the individual sets in the ensemble.

Aksela [7.4]) compared six different methods to select the members based on six diversity measures among them: (i) Correlation between errors (ii) Q-statistics (iii) Mutual information (iv) Ratio between different and same errors (v) Weighted count of errors and correct results (vi) Exponential error count. The results favor the exponential error count method, which weights identical errors made by classifiers in an exponential fashion and normalize it with the number of data points where all members are correct.

Navone et al [7.29] proposed an incremental approach in which at every stage, the system searched for a new member that is partially anti-correlated with the current members in the ensemble. However, in their paper [7.29], diversity is not directly resulting from their method but is rather the theoretical result from the assumption that the added member has to reduce the ensemble error. It was not clear if this theoretical diversity did exist in the ensemble, as most of other papers claiming the use of diversity related techniques.

Banfield et al [7.6] defined another diversity measure - Percentage Correct Diversity Measure - which counts the number of examples that between 10% to 90% of the classifiers correctly classify a data instance. Also, they proposed two different diversity based mechanisms to thin out the classifiers in the pool ensemble. However, the used classifiers in their research were CA's instead of neural network.

Oliveira et al [7.30] introduced the use of MOP to two levels of ensemble classification : feature selection and member selection. The argument is that Pareto optimality displays characteristics suitable for diversity and ensemble [7.1, 7.23, 7.30]. The second level includes selecting the subset of networks that promotes maximization of recognition rate and maximization of ambiguity (diversity) by using a Pareto-based approach. Although the method seems to perform well, there is not any validation for the claim that the members are diverse.

Both [7.30] and [7.2] used MOP as the main mechanism to promote diversity among the members of the ensemble. However, because of the crisp comparison being used in MOP, the networks could be similar and MOP could still pick up the subset with highest diversity measure, though "highest" may mean very little difference from the lesser values, e.g. $3.5 > 3.4999$.

Lazarevic and Obradovic [7.22] derived their pruning method based on unsupervised clustering algorithm where the distance between clusters represents the diversity between classifiers. The algorithm incrementally increases the number of clusters until the diversity between clusters deteriorates then the clusters are removed, leaving only their centroids.

7.3 Comparisons of Different Ensemble Methods

In this section, we will discuss experiments with different setups to gain a clearer understanding on the important aspects that may influence the performance of an ensemble.

The investigation focuses on six group of methods, namely (i) simple evolutionary computation with no ensemble (*i.e.* the best network found all over an evolutionary run is selected), (ii) memetic, through back propagation, evolutionary computation with no ensemble, (iii) ensembles built using simple evolutionary computation, (iv) ensembles built using an island model, (v) ensemble built using evolutionary computation with negative correlation learning, and (vi) ensembles built using evolutionary Pareto multi-objective optimization. Beside exploring the previous methods, we will also use two generalization improvement methods: noise injection and early stopping criteria. The first generalization improvement method adds noise in the evolution with the aim of introducing stochastic noise to the fitness landscape hoping that the evolutionary method will escape a local optima. The second generalization improvement method uses some criteria, such as performance on a validation set, to select intermediate networks in the evolutionary run. These networks may not be the best found network in terms of training error but may have the potential to generalize better.

The rest of the chapter is divided into three sub-sections. The "Method" sub-section presents a detailed description of each method to be used in the experiments. The "Experimental set-up" sub-section presents the objectives of the experiments, a description of the five data set being used, and the choice of parameters for different methods.The "Results and analysis" subsection provides a detailed analysis to the results.

7.3.1 Methods

Ensemble methods. An ensemble [7.41] is a set of redundant networks, each by itself would "provide a solution to the same task", i.e. predict the target output of an input vector. Thus, the output of an ensemble could be computed by feeding the input pattern to the individual neural nets, getting a prediction from each, and combining their outputs in a predefined fashion, namely a combination gate.

Self-adaptive evolutionary strategies ($\mu + \lambda$). The Self-Adaptive Evolutionary Strategies is constructed based on the log-normal self–adaptation, where a rotation angle is used to adapt the search towards coordinates which are likely to be correlated [7.40]. In this algorithm, the children are created by discretely combining and mutating a pair of parents using the self adaptive step size σ.

Algorithm 1 *The self–adaptive evolutionary strategy ($\mu + \lambda$).*

Step 0. Randomly generate μ parents, where each parent $z_k = (\boldsymbol{x}_k, \boldsymbol{\sigma}_k)$.

Step 1. Set $\tau = \left(\sqrt{\left(2\sqrt{(n)} \right)} \right)^{-1}$ and $\tau' = \left(\sqrt{(2n)} \right)^{-1}$.

Step 2. Until the halting criteria are satisfied do

  Step 2.1. Until λ children are generated, do

    Step 2.1.1. Select two parents $z_k = (\boldsymbol{x}_k, \boldsymbol{\sigma}_k)$ and $z_l = (\boldsymbol{x}_l, \boldsymbol{\sigma}_l)$ at random to generate child $\boldsymbol{y}_j = (\boldsymbol{x}_j, \boldsymbol{\sigma}_j)$.

    Step 2.1.2. Discrete recombination: for each variable x_{ji} and step size σ_{ji} in $\boldsymbol{y}_j$, do $(x_{ji} = x_{ki}$ and $\sigma_{ji} = \sigma_{ki})$ or $(x_{ji} = x_{li}$ and $\sigma_{ji} = \sigma_{li})$.

    Step 2.1.3. Mutation: For each x_{ji} and step size σ_{ji} in $\boldsymbol{y}_j$

$$x'_{ji} = x_{ji} + \sigma_{ji} N_j(0,1). \tag{7.2}$$

$$\sigma'_{ji} = \sigma_{ji} \exp(\tau' N(0,1) + \tau N_j(0,1)). \tag{7.3}$$

  Step 2.2. Select the best μ individuals among all the $\mu + \lambda$ parents and children.

Neuro-ensemble using the island model. The Parallelizing Island model [7.3, 7.5, 7.24] is one type of diversity-promotion mechanisms often used for distributing the computation in the evolutionary algorithm. The main difference between island model and a normal evolutionary computation method is the division of the population into sub-populations of number of isolated islands. The sub-population occupying each island is updated independent from other sub-populations. Occasionally, members in neighboring islands are migrated from one island to another. In one variant of the algorithm, the best individuals from one island replace the worst individuals of neighboring islands. The connections between islands can be random or fixed in a sequential and cyclic manner.

Neuro-ensemble using negative correlation learning. Negative Correlation Learning, proposed by Yao and Liu [7.26, 7.27], aims to maximize the difference among the members of the ensemble by adding a penalty term in each member network's mean squared error (MSE) function. These penalty terms are computed based on the correlations of the individuals. The networks are trained using normal Back Propagation [7.18].

Let M be the total number of members in the ensemble (often termed ensemble size). For each pattern X^p in the training set, the output of the ensemble F^p is computed as the average of the output $\hat{Y}^p(m)$, $m = 1 \ldots M$ of the M individual members.

$$F^p = \frac{1}{M} \sum_{m=1}^{M} \hat{Y}^p(m). \tag{7.4}$$

The error function for network m is defined by

$$Error^p(m) = \frac{1}{P} \sum_{p=1}^{P} \frac{1}{2}(\hat{Y}^p(m) - Y^p)^2 + \frac{1}{P} \sum_{p=1}^{P} \lambda \Phi^p(m). \tag{7.5}$$

$\Phi^p(m)$ is called the penalty function of network m and pattern p. This represents the correlation between the networks.

$$\Phi^p(m) = (\hat{Y}^p(m) - F^p) \sum_{l \neq m} (\hat{Y}^p(l) - F^p). \tag{7.6}$$

The partial derivative of the error $Error^p(m)$ with respect to the output of network m is

$$\frac{\partial Error^p(m)}{\partial \hat{Y}^p(m)} = (\hat{Y}^p(m) - Y^p) - \lambda(\hat{Y}^p(m) - F^p). \tag{7.7}$$

In other words, the only difference between Negative Correlation Learning and BP is an additional penalty term of $\lambda(\hat{Y}^p(m) - F^p)$ to the error function.

7.3.2 Evolutionary Multi Objective Optimization Application in Ensemble

Pareto-based methods are potential diversity mechanisms [7.2, 7.23]. In this set of experiments, we investigate the feasibility of using Pareto ranking as a source of diversity for the ensemble. In the Multi Objective Optimization literature, diversity has been also one of the main concerns. As a result, a number of methods [7.12, 7.47, 7.57] were invented to solve this problem. A state-of-the-art method,Strength Pareto Evolutionary Algorithm (SPEA), was proposed by Zitzler [7.57], where the Pareto ranking algorithm was introduced as a diversity mechanism. The algorithm takes into account the crowdness (based on a special dominance/nondominance ratio) of the population in different regions and promote diversity by avoiding crowded regions. This ranking method results in a wide spread of solutions on the Pareto frontier. To be fair in our comparison, Zitzler's Pareto ranking algorithm is implemented together with the Evolutionary Strategy algorithm introduced earlier.

Algorithm 7.3.2 shows Zitzler's Pareto ranking method, in which an elitist set P' is created to hold the accounted non-dominated solutions in the whole population P. For each member i of the elitist set, a strength function (i.e. crowding fitness) is defined as the number of individuals in the whole population P being dominated by i, normalized by N, the total number of individuals in the whole population. Hence, the fitness of a non-dominated individual is always less than 1. Next, for each dominated member j in the population P, a fitness function is computed as the sum of the strength of the elitist members that dominate j. To ensure that the non-dominated solutions have better chance (i.e. lower crowding fitness) to survive, a value of 1 is added to the fitness of the dominated solutions.

Algorithm 2 *Pareto ranking algorithm.*

Input:

– *Population P*
– *Population size N*

Step 1. Create an empty pool P'
Step 2. Copy all non-dominated individuals into P'
Step 3. For each non-dominated solution i in P' , define a strength function

$$s(i) = \frac{N_1}{N+1}. \tag{7.8}$$

where N_1 is the number of individuals in P that are dominated by i. Assign a fitness $f(i) = s(i)$ to item i
Step 4: For each dominated solution in $P - P'$, assign a fitness $f(i) = 1 + \sum_{j \in P', j\, dominates\, i} s(j)$
Step 5: Rank the population in ascending order of the fitness

---

The Pareto ensemble method uses Pareto ranking as the diversity promoting mechanism. It is similar to the evolutionary ensemble algorithm with the only difference lies in the use of Pareto ranking as the mechanism for selection. The training set is split into two subsets and the multi-objective problem is then to minimize the training error on each subset.

Overfitting avoidance methods.

Noise injection. A random Gaussian noise with zero mean and a standard deviation of σ is added to the training error (fitness) of the ensemble in each generation.

Stopping criteria. In this chapter, we will investigate three different criteria for choosing the optimal generation for the ensemble. The first criterion is to use the population of the last generation to form the ensemble; this is widely used in the literature [7.27, 7.51, 7.53]. The second criterion is to form the ensemble in each generation and evaluate it on the validation set. The ensemble corresponding to the minimum validation error is selected. The third criterion is to form the ensemble from the members of the population achieving the minimum average fitness on the validation set.

Gating methods for combining members in the ensemble. There are many ways to combine the individual ANNs to form the ensemble. In our experiment, we will exploit the three most popular combination (gating) methods in the ensemble literature: majority voting, average and winner-take-all.

Majority voting In this combination method, the output of the ensemble will be the binary output value which receives the vote of the majority of the networks in the ensemble

Winner-take-all In the winner-take-all method, the ensemble output is the output of the network whose activation is maximally different from the classification threshold (e.g. 0.5).

$$y_{ens}(\vec{x}_i) = \arg_{max,j}(|\hat{y}_j(\vec{x}_i) - threshold|) j \in [1, N_c]. \tag{7.9}$$

Simple averaging In the average method, the output of the ensemble is the mean of the output activations of the networks.

$$y_{ens}(\vec{x}_i) = \sum_{j=1}^{N_c} \hat{y}_j(\vec{x}_i)/N_c. \tag{7.10}$$

Diversity measure. In this initial investigation, we will apply a simple diversity measure based on the average hamming distance of the predicted outputs of all possible pairs of individuals in the networks. In other words, let N be the number of instances in the training set, M the number of classifiers in the ensemble, and $\vec{y}_m = y_m^i, i = 1..N, m = 1..M$, where each entry y_m^i corresponds to the predicted class of example i in the training set for classifier m. Then, the diversity measure for two vectors $\vec{y}_j$ and $\vec{y}_k$ is

$$d_{jk} = \sum_{i=1}^{N}(a(y_j^i, y_k^i)). \tag{7.11}$$

where

$$a(y_j^i, y_k^i) = 0 \quad if\, y_j^i = y_k^i, \quad 1 \quad otherwise. \tag{7.12}$$

Finally, the diversity measure of the ensemble is

$$D_{ens} = \sum_{p=1}^{M} \sum_{q=1, q<>p}^{M} (d_{pq}). \tag{7.13}$$

7.3.3 Experiment Set-up

Hypotheses. The experiments are designed around seven hypotheses; these are:

Hypothesis 1: Combination of local search through learning and evolution improves the generalization ability

The first set of experiments is designed to analyze the effect of combining learning with evolution. In the literature of evolutionary artificial neural networks, it is still debatable if combining learning and evolution will be better than each alone.

Hypothesis 2: an ensemble of neural networks perform better than individual networks

The second set of experiments is used to verify the claim by many researchers [7.19, 7.25, 7.26, 7.27, 7.37, 7.41, 7.45, 7.46, 7.51, 7.52, 7.53, 7.55] that ensemble of classifiers performs better than individual ones.

Hypothesis 3: different combination methods yield similar results

In this set of experiments, the members of the ensemble are combined using different combination gates. We would like to verify the effect of these different combination gates.

Hypothesis 4: Noise injection improves generalization ability by changing the bias/variance distribution

Noise could be injected in different levels of the system. In this experiment, we chose to inject a random Gaussian noise to training fitness of the networks in the ensemble with the hope that it could reduce the pressure of evolution to overfit the networks. This disturbance gives networks with bad training fitness more chance to survive because often the ones with less ability to memorize the training data are the ones which can generalize well. This noise addition also has the effect of changing the bias/variance distribution of each individual network [7.8].

Hypothesis 5: Early stopping is useful in avoiding overfitting

The next set of experiments is motivated by the observation frequently seen in the neural network literature that early stopping could reduce the chance of overfiting and thus improving the network's generalization ability. However, in the literature of ensemble, it is not yet clear if different stopping criteria affect the generalization ability of the ensemble. Therefore, this experiment is designed to explore possible use of early stopping and possible stopping criteria.

Hypothesis 6: Useful diversity is an important feature of ensemble techniques

Diversity, as seen in the literature review, is a very interesting and important issue in the ensemble. However, most of the papers in the field don't report the actual diversity of the members in their ensembles. In this initial investigation, three different diversity mechanisms, namely negative correlation learning, island model and the Pareto-based bootstrapping model, are examined in two aspects: accuracy and diversity.

The aim here is not to claim if any method outperforms the other. Since there are many different elements that could affect greatly the performance of a method, and since the aim of this investigation is to gain an understanding on different mechanisms, we choose a common evolutionary method with a common parameter setting for all three mechanisms. There is no guarantee that the chosen evolutionary method and parameters are optimized for each mechanism.

Hypothesis 7: Architecture complexity is important in the problem of over-fitting

It is widely noted that architecture complexity plays an important role in the generalization ability of neural networks. Small simple networks have less chance to memorize the training data and, thus, have more bias and less variance. On the other hand, large and complex networks could capture more of the data but run into the problem of overfitting the data. Hence, the problem of finding a suitable architecture for each problem set is a difficult problem. In this experiment, the architecture is changing by changing the number of hidden units in the three-layer feed-forward neural networks.

Data sets. The hypotheses are tested on five standard data sets taken from the UCI Machine Learning Repository: (i) the breast cancer Wisconsin, the Australian credit card assessment, the diabetes, the liver disorder and the tic-tac-toe. These data sets were downloaded from ice.uci.edu [7.7].

The Breast Cancer data set was originally obtained from W. HG. Wolberg at the University of Wisconsin Hospitals, Madison. The set is divided into two classes: benign or malignant. The set has 699 instances (as of 15 July 1992) represented by 9 categorical attributes. 458 instances are benign and 241 are malignant.

The Diabetes data set was donated by Vincent Sigillito from Johns Hopkins University and was constructed by constrained selection from a larger database by the National Institute of Diabetes and Digestive and Kidney Diseases. All patients here are females at least 21 years old of Pima Indian heritage. The set consists of 768 instances of 8 attributes. 500 examples tested negative for diabetes, and 268 are tested positive.

The liver disorder data set was donated by Richard S. Forsyth from the collected data by BUPA Medical Research Ltd. The set has 345 instances with 6 attributes. The class distribution is 145 (class 0) and 200 (class 1).

The Australian credit card assessment data set contains 690 instances of 15 attributes. There are two classes: + with 307 instances and − with 383 instances.

The Tic-Tac-Toe endgame data set was created and donated by David W. Aha on 19 August 1991. This database encodes the complete set of possible board configurations at the end of tic-tac-toe games, where "x" is assumed to have played first. The target concept is "win for x" (i.e., true when "x" has one of 8 possible ways to create a "three-in-a-row"). There are 958 instances of 9 attributes (each corresponding to one tic-tac-toe square). There are two classes: negative (332 instances) and positive (625 instances).

In the experiments, we used ten-fold cross validation for each data set. A data set is divided into ten subsets using the stratified sampling method. For each fold, the distribution of data in the test set, the validation set and the training set is 1:1:8. For each fold, a different prefixed random seed is used to generate the required random numbers (e.g. network weights, noise, crossover and mutation rates) for the method. The method is trained using the training set, stopped by one of the different criteria using the validation set, and the ensemble obtained by combining the population at the stopping point is tested on the test set. The final result is the average of the ten-fold results.

Parameters. Each individual feed-forward neural network consists of a single hidden layer with 1-6 hidden units. The evolution is run for a maximum of 1000 generations, unless an early stopping criterion is satisfied, with a population of ($\mu = 20$ and $\lambda = 80$) for all methods except the island model and 10 islands of ($\mu' = 2$ and $\lambda' = 8$) for the island model. The migration interval in the island model method is set to 10 generations.

For the negative correlation learning method, we used a learning rate of 0.1 for BP with 10 epochs of learning. For the negative correlation penalty coefficient, we also tested different values ranging from 0.1 to 0.5. The initial experiment showed that 0.2 is a suitable value for most of the data sets, and thus, it is used in the NCL based experiments.

For the Ensemble using MOP, the two objectives are the training errors of the neural network on two disjointed subsets of the training data.

Finally, for the ensemble method using noise distortion, a Gaussian noise of zero mean and 0.01 standard deviation is applied on the computed training fitness of each individual neural network.

7.3.4 Results and Analysis

The experiments, as discussed above, are carried out on six group of methods: (i) the simple evolutionary computation, (ii) the memetic (through back propagation) evolutionary computation, (iii) the ensemble of simple evolutionary computation, (iv) the ensemble of evolutionary neural network using the island model, (v) the ensemble of evolutionary negatively correlated neural networks, and (vi) the ensemble of evolutionary Pareto based neural networks. Each group consists of a with-noise and a without-noise experiments. The means and standard deviations of the testing errors in these twelve experiments are recorded based on the three stopping criteria. Obviously, there are a number of different dimensions to be considered such as the use of local search, the effect of noise injection, the various architecture complexity through the number of hidden units, and different diversity mechanisms. The large amount of experimental results are grouped under the seven hypotheses. To verify a hypothesis, a minimal error value is taken across all the average of all other dimensions.

Local search helps evolution to find better solutions. Table 7.1 shows the results of the first set of experiments, in which a simple evolutionary computation (EC) approach is used to evolve, with and without Back Propagation (BP) learning, the best neural network to classify the five data sets. These results show that integrating Back Propagation to EC improves the performance of three out of the five data sets. The performance enhancement of the Liver Disorder (16%) and Australian Credit Card (11.2%) Data sets and the performance worsening of Breast Cancer Wisconsin Data set (13.3%) are significant. The results in Table 7.1 implies that Back Propagation learning, on the average, helps EC to find better solutions for the classification problem.

Table 7.1. Memetic effect - evolutionary ANN with and without back propagation learning

Data set	EC	EC + BP learning
Breast Cancer Wisconsin	**0.030(0.017)**	0.034(0.021)
Australian Credit Card	0.150(0.043)	**0.134(0.041)**
Diabetes	0.242 (0.055)	**0.231 (0.044)**
Liver Disorder	0.374 (0.054)	**0.318 (0.085)**
Tic Tac Toe Games	0.289 (0.041)	0.291 (0.049)

Ensemble performs better than individuals. The second set of experiments compares the simple ensemble with the best individual (without ensemble) approaches. The results are summarized in Table 7.2. The results show that Simple Ensemble method performs better than the best individual in three out of five data sets, especially the improvement in the Breast Cancer and the Liver Diorder data sets are quite significant (**10.1% and 10.7%**). In the only data set (Tic Tac Toe) where the best individual outperforms the ensemble, the improvement is actually quite small (**3.6%**). As a result, ensemble of neural networks on the average performs better than the best neural network in the population.

Table 7.2. Ensemble vs. best individual

Data set	Best Individual	Simple Ensemble
Breast Cancer Wisconsin	0.030(0.017)	**0.027 (0.014)**
Australian Credit Card	0.150(0.043)	**0.140 (0.052)**
Diabetes	0.242(0.055)	0.242(0.056)
Liver Disorder	0.374(0.054)	**0.336(0.075)**
Tic Tac Toe Games	**0.289(0.041)**	0.299(0.052)

Combination gates perform similar. Table 7.3 show that the performance of different gates is similar. This suggests that on the average, the three different gates perform the same.

Table 7.3. Comparison of three different combining gates: (i) majority voting, (ii) averaging, (iii) winner-take-all

Data set	Majority voting	Averaging	Winner-take-all
Breast Cancer Wisconsin	0.027(0.014)	0.027(0.017)	0.027(0.017)
Australian Credit Card	0.140(0.050)	**0.133(0.048)**	0.139(0.050)
Diabetes	0.238(0.043)	0.235(0.041)	**0.232(0.037)**
Liver Disorder	0.286(0.038)	0.286(0.038)	0.286(0.038)
Tic Tac Toe Games	**0.212(0.042)**	**0.212(0.042)**	0.223(0.044)

Noise injection improves generalization. Table 7.4 shows the results of the comparison of with and without noise disturbance in each data set. Except in the Breast Cancer Wisconsin Data set, where three out of five methods perform better without noise addition, the remaining four data sets exhibit a clear preference for noise injection. Also, from Table 7.4, noise intrusion is favorable for Simple Ensemble and Ensemble with Multi-Objective Optimization methods where four out of five data sets display preference on noise.

In summary, the results from Table 7.4 suggest that the performance could be improved by introducing a Gaussian noise into the fitness of the neural network during fitness computation. Especially, noise addition works well with Simple Ensemble and Ensemble with MOP methods. A possible reason is that noise addition reduces the pressure of selecting overly-fit network using the training fitness.

Early stopping is useful in avoiding overfitting. Table 7.5 displays the results of the set of experiments to verify hypothesis 5: early stopping performs better than stopping at the maximum number of generations. As seen from the outcomes, using two early stopping criteria: minimum of the ensemble on validation and minimum average of population fitness on validation outperforms (4 of 5 data sets per method show preference) the last generation criterion for "Simple Ensemble", "Ensemble with Island model", and "Ensemble with Negative Correlation Learning". Only "Ensemble with MOP" shows a slight favor of the last generation criterion. The results support the hypothesis that early stopping is beneficial.

Useful diversity is an important feature of ensemble techniques. Table 7.6 displays the results of classification of five data sets using four different ensemble methods: Simple Ensemble, Ensemble with Island model, Ensemble with NCL, and Ensemble with MOP. The values show that Island model performs the best in three data sets and Negative Correlation Learning performs the best in the remaining two data sets.

Table 7.4. Effect of noise injection for each data set

Breast Cancer		
Method	Without Noise	With Noise
EC	**0.030(0.017)**	0.032(0.016)
Simple Ensemble	**0.027(0.014)**	0.028(0.019)
Ensemble w/ island	0.027(0.017)	**0.025(0.017)**
Ensemble w/NCL	**0.027(0.017)**	0.030(0.017)
Ensemble w/MOP	0.030(0.018)	**0.027(0.017)**

Australian Credit Card Assessment		
Method	Without Noise	With Noise
EC	0.150(0.043)	**0.142(0.048)**
Simple Ensemble	0.140(0.052)	**0.137(0.044)**
Ensemble w/ island	**0.133(0.048)**	0.149(0.043)
Ensemble w/NCL	0.144(0.040)	**0.137(0.044)**
Ensemble w/MOP	**0.140(0.050)**	0.144(0.057)

Diabetes		
Method	Without Noise	With Noise
EC	0.242(0.055)	**0.235(0.045)**
Simple Ensemble	0.242(0.056)	**0.240(0.060)**
Ensemble w/ island	**0.232(0.037)**	0.246(0.055)
Ensemble w/NCL	0.236(0.050)	**0.226(0.057)**
Ensemble w/MOP	0.236(0.051)	**0.226(0.042)**

Liver Disorder		
Method	Without Noise	With Noise
EC	0.374(0.054)	**0.321(0.054)**
Simple Ensemble	0.336(0.075)	**0.315(0.068)**
Ensemble w/ island	0.350(0.085)	**0.303(0.080)**
Ensemble w/NCL	**0.286(0.038)**	0.295(0.054)
Ensemble w/MOP	0.326(0.074)	**0.298(0.073)**

Tic Tac Toe Games		
Method	Without Noise	With Noise
EC	**0.289(0.041)**	0.322(0.047)
Simple Ensemble	0.299(0.052)	**0.281(0.062)**
Ensemble w/ island	0.311(0.034)	**0.277(0.042)**
Ensemble w/NCL	0.212(0.042)	**0.195(0.039)**
Ensemble w/MOP	0.278(0.042)	**0.272(0.026)**

Table 7.5. Comparison of different stopping criteria (i) last generation (ii) minimum on validation (iii) minimum average of population on validation

Simple Ensemble			
Data set	Last generation	Minimum on validation	minimum average
Breast Cancer Wisconsin	0.035(0.028)	**0.027(0.014)**	0.030(0.021)
Australian Credit Card	0.149(0.056)	**0.140(0.059)**	**0.140(0.052)**
Diabetes	0.243(0.047)	**0.242(0.056)**	0.248(0.039)
Liver Disorder	0.338(0.066)	0.347(0.071)	**0.336(0.075)**
Tic Tac Toe Games	**0.299(0.052)**	0.308(0.049)	0.301(0.053)

Ensemble with Island model			
Data set	Last generation	Minimum on validation	minimum average
Breast Cancer Wisconsin	0.219(0.246)	0.030(0.012)	**0.027(0.017)**
Australian Credit Card	0.150(0.057)	**0.133(0.048)**	0.142(0.051)
Diabetes	0.244(0.041)	**0.232(0.037)**	0.235(0.038)
Liver Disorder	0.422(0.071)	**0.350(0.085)**	0.368(0.064)
Tic Tac Toe Games	**0.311(0.034)**	0.317(0.047)	0.323(0.034)

Ensemble with Negative Correlation Learning			
Data set	Last generation	Minimum on validation	minimum average
Breast Cancer Wisconsin	0.044(0.028)	0.028(0.019)	**0.027(0.017)**
Australian Credit Card	0.169(0.048)	0.147(0.044)	**0.144(0.040)**
Diabetes	0.264(0.058)	**0.236(0.050)**	0.249(0.051)
Liver Disorder	**0.286(0.038)**	0.292(0.071)	0.295(0.071)
Tic Tac Toe Games	0.217(0.038)	0.224(0.040)	**0.212(0.042)**

Ensemble with Multi Objective Optimization			
Data set	Last generation	Minimum on validation	minimum average
Breast Cancer Wisconsin	0.032(0.020)	0.032(0.017)	**0.030(0.018)**
Australian Credit Card	**0.140(0.050)**	0.146(0.052)	0.142(0.044)
Diabetes	0.248(0.057)	**0.236(0.045)**	0.246(0.052)
Liver Disorder	**0.326(0.074)**	0.347(0.078)	0.347(0.080)
Tic Tac Toe Games	**0.278(0.042)**	0.293(0.052)	0.284(0.050)

Table 7.6. Ensemble without vs. with diversity promotion mechanism

Breast Cancer Wisconsin			
Simple Ensemble	Ensemble + Island	Ensemble + NCL	Ensemble + MOP
0.027(0.014)	**0.027(0.017)**	0.027(0.017)	0.030(0.018)

Australian Credit Card			
Simple Ensemble	Ensemble + Island	Ensemble + NCL	Ensemble + MOP
0.140(0.052)	**0.133(0.048)**	0.144(0.040)	0.140(0.050)

Diabetes			
Simple Ensemble	Ensemble + Island	Ensemble + NCL	Ensemble + MOP
0.242(0.056)	**0.232(0.037)**	0.236(0.050)	0.236(0.045)

Liver Disorder			
Simple Ensemble	Ensemble + Island	Ensemble + NCL	Ensemble + MOP
0.336(0.075)	0.350(0.085)	**0.286(0.038)**	0.326(0.074)

Tic Tac Toe Games			
Simple Ensemble	Ensemble + Island	Ensemble + NCL	Ensemble + MOP
0.299(0.052)	0.311(0.034)	**0.212(0.042)**	0.278(0.042)

The next question is whether these diversity mechanisms do generate diversity. Table 7.7 shows the diversity measures of five data sets across different stopping criteria and diversity methods. The bold values are the high diversity measure (members more disagree) in contrast to plain numbers which correspond to little difference among ensemble members. Moreover, Figs. 7.1, 7.2, 7.3, 7.4, and 7.5 plot the Diversity over the generations of the four different diversity mechanisms for the five Data sets.

Looking across different ensemble methods with different diversity promoting mechanisms, ensembles generated with the Island model show very good diversity level for most cases, with "Simple Ensemble" and "Ensemble with Multi Objective Optimization" are reasonably good in promoting diversity. However, surprisingly, "Ensemble with Negative Correlation Learning" shows a very poor diversity measure. This fact is supported by the analysis by McKay and Abbass [7.28] which proved that Negative Correlation Learning does not promote diversity among member networks.

Considering the four lines in each Fig. (7.1, 7.2, 7.3, 7.4, and 7.5), which correspond to four different methods, "Ensemble with Island model" and "Ensemble with MOP" have the best diversity curves (top two lines in each graph), and the "Ensemble with Negative Correlation" Curves are very low for all data set. These results agrees with the remark that the former two methods produce higher diversity among members than the other methods. Also, it confirms that Negative Correlation Learning produces very small diversity among the members.

Combining the diversity analysis and performance analysis across different methods, we came up with a contradicting fact: Island model yields

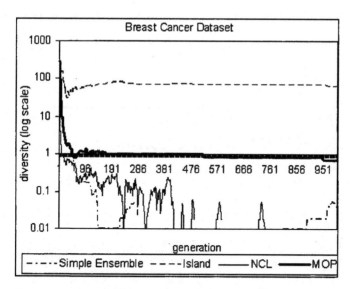

Fig. 7.1. Diversity measure of the breast cancer wisconsin data set

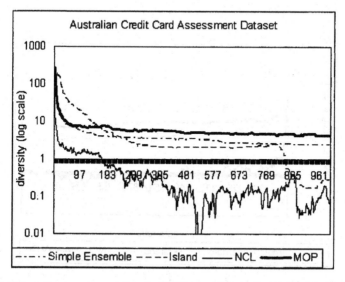

Fig. 7.2. Diversity measure of the Australian credit card assessment data set

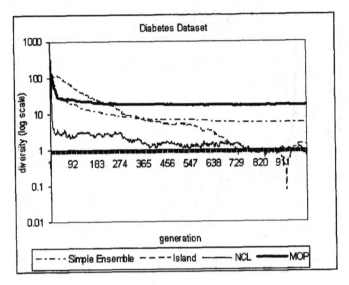

Fig. 7.3. Diversity measure of the diabetes data set

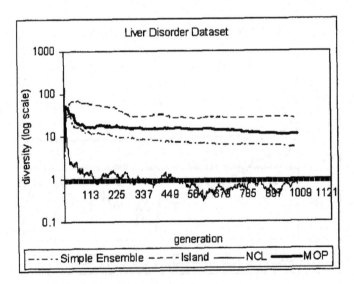

Fig. 7.4. Diversity measure of the liver data set

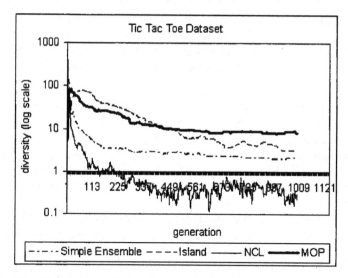

Fig. 7.5. Diversity measure of the Tic Tac Toe games Data set

best performance and the highest diversity level, while Negative Correlation Learning also performs well but has very small diversity. Also, simple ensemble and ensemble with MOP has quite good diversity levels but with poor performance. In conclusion, we have not found any useful connection between diversity and performance when building an ensemble.

One possible reason is that unhealthy diversity could deteriorate the ensemble performance. Let us re-visit Krogh's equation:

$$E = \overline{E} - \overline{A}. \tag{7.14}$$

Promoting diversity among members of the ensemble will change the Ambiguity term $\overline{A}$ which contains all the correlations among the individual classifiers. However, large diversity in the ensemble may greatly reduce the average accuracy of the ensemble $\overline{E}$. Thus, it is possible that unhealthy diversity, which reduces $\overline{E}$ much faster than $\overline{A}$, can degrade the performance of the ensemble. In the future, more experiments are required to fully analyze the relationship of diversity and accuracy.

Diversity across different stopping criteria. Looking across the columns in Table 7.7, the minimum on validation (column 2) stopping criterion shows reasonably high level of diversity across three stopping criteria. The second best is the minimum average of population on validation (column 3) and the lowest is the last generation criterion. This result, together with the previous conclusion that early stopping performs better than last generation, suggests a connection between high diversity and the good performance of early stopping.

Also, Figs. 7.1, 7.2, 7.3, 7.4, and 7.5 show the diversity decaying over time (Note: since the diversity measure decreases quickly to zero, the graph is plotted in log scale). From the figures, since diversity tends to reduce over time, an early stopping often implies higher diversity among the individuals.

Architecture complexity is important. This final set of experiments is designed to testify any connection of the neural network's architecture complexity and the performance of the ensemble. Table 7.8 presents the results of different number of hidden units on the performance of the ensemble for three data sets: the Australian credit card, the Diabetes and the Tic Tac Toe. Looking across the three data sets, it is obvious that there are no particular fixed number of hidden units that can perform best for all data sets with all methods.

Looking down the columns in Table 7.8, island model appears to perform comparably well across various number of hidden units while NCL favors less complex networks and MOP prefers more complex ones.

The results in Table 7.8 back the claim that architecture complexity of the neural network is an important factor to be considered when designing an ensemble method. Moreover, and what is more interesting, the level of network complexity depends on the diversity promoting mechanism.

7.4 Conclusion

In this chapter, we have presented an overview of some state-of-the-art methods in the field of Ensemble of Artificial Neural Networks. A number of experiments were conducted to investigate various aspects of methods used to build neuro-ensembles. The findings verify some points raised by other researchers in the field, and also raise a number of interesting and potential directions for future research.

The key points from the experimental analysis could be summarized as : (1) Different combination gates have little effect on the performance of the ensemble, (2) The diversity level maintained by negative correlation learning is poor, which was suggested by McKay and Abbass's analysis [7.28], (3) Combining individuals (Ensemble) improves the performance of the system.

Some interesting open directions are raised from the experimental results. Those are: (1) Local search does help evolution to find better solutions, this is applicable for both ensemble and individual-based methods, (2) Noise injection shows interesting effects on performance enhancement, though the improvement is not yet clear from our initial experiments, (3) Early stopping is useful in enhancing generalization, where using different minimum values in the validation fitness, such as the fitness of the ensemble or the average fitness of the population, can avoid the networks to overfit the training data, (4) Architecture complexity of the neural networks is an important factor

Table 7.7. Diversity measure of five Data sets across different diversity mechanisms and stopping criteria

	Breast Cancer Wiscosin		
Objective	last generation	min ensemble on validation	min average of population
Simple Ensemble	0.039(0.118)	**28.042(20.423)**	3.203(2.730)
Ensemble Island	**59.912(57.168)**	**91.205(41.182)**	**22.853(20.531)**
Ensemble NCL	0.000(0.000)	**116.646(139.802)**	1.542(1.342)
Ensemble MOP	0.721(1.727)	**56.438(63.541)**	4.009(4.815)

	Australian Credit Card Assessment		
Objective	last generation	min ensemble on validation	min average of population
Simple Ensemble	2.381(2.537)	**78.141(42.795)**	**20.452(22.799)**
Ensemble Island	0.285(0.857)	**85.813(58.874)**	**16.838(15.241)**
Ensemble NCL	0.060(0.151)	**31.515(83.813)**	3.600(4.638)
Ensemble MOP	4.478(3.769)	**48.713(30.609)**	12.235(9.887)

	Diabetes		
Objective	last generation	min ensemble on validation	min average of population
Simple Ensemble	5.167(3.165)	**51.202(27.868)**	**18.474(19.099)**
Ensemble Island	1.282(3.600)	**64.383(41.064)**	**28.635(22.122)**
Ensemble NCL	0.743(0.528)	6.460(9.215)	3.552(2.457)
Ensemble MOP	16.985(10.014)	**34.303(15.857)**	**25.267(10.204)**

	Liver Disorder		
Objective	last generation	min ensemble on validation	min average of population
Simple Ensemble	5.360(4.938)	**23.387(17.791)**	**16.083(19.161)**
Ensemble Island	**24.479(12.618)**	**62.334(12.630)**	**44.370(19.247)**
Ensemble NCL	0.734(0.938)	2.415(1.185)	1.681(1.208)
Ensemble MOP	10.717(7.366)	**44.920(36.224)**	**23.461(13.196)**

	Tic Tac Toe Game		
Objective	last generation	min ensemble on validation	min average of population
Simple Ensemble	2.048(2.970)	8.608(13.847)	3.836(4.208)
Ensemble Island	2.977(4.559)	**59.845(32.055)**	**37.045(46.293)**
Ensemble NCL	0.215(0.209)	3.856(2.879)	1.815(2.509)
Ensemble MOP	8.173(10.266)	**39.862(25.063)**	**18.239(19.456)**

Table 7.8. Various architecture complexity by varying number of hidden units

Australian Credit Card Assessment			
Hiddens	Island model	NCL	MOP
1	**0.134(0.054)**	**0.133(0.044)**	0.152(0.035)
2	0.147(0.047)	**0.133(0.050)**	0.146(0.040)
3	0.139(0.052)	0.139(0.049)	0.143(0.045)
4	0.150(0.045)	0.144(0.048)	0.142(0.049)
5	0.149(0.051)	**0.133(0.039)**	0.144(0.057)
6	**0.133(0.048)**	0.144(0.040)	**0.139(0.048)**

Diabetes			
Hiddens	Island model	NCL	MOP
1	0.246(0.046)	**0.225(0.056)**	0.246(0.055)
2	**0.236(0.050)**	**0.229(0.062)**	**0.235(0.074)**
3	0.244(0.061)	**0.226(0.069)**	0.247(0.044)
4	**0.234(0.058)**	0.230(0.063)	0.242(0.059)
5	**0.232(0.042)**	0.231(0.068)	**0.238(0.056)**
6	**0.232(0.037)**	0.236(0.050)	**0.236(0.051)**

Tic Tac Toe			
Hiddens	Island model	NCL	MOP
1	0.318(0.015)	0.305(0.057)	0.311(0.050)
2	0.322(0.044)	0.309(0.045)	**0.261(0.045)**
3	0.312(0.046)	0.286(0.033)	0.313(0.055)
4	**0.300(0.048)**	**0.249(0.034)**	0.297(0.030)
5	0.313(0.037)	**0.241(0.071)**	0.302(0.047)
6	**0.311(0.034)**	**0.212(0.042)**	**0.279(0.042)**

to be considered when designing the ensemble of EANNs, (5) The connection between diversity and performance of the ensemble is still a myth to be verified, this is supported by the results of Kuncheva's experiments [7.23].

References

7.1 Abbass, H. A. (2002): An evolutionary artificial neural networks approach for breast cancer diagnosis. Artificial Intelligence in Medicine, **25**, 265–281

7.2 Abbass, H. A. (2003): Formation of neuro-ensembles using the pareto concept, to appear

7.3 Adamidis, P. (1994): Review of parallel genetic algorithms bibliography. Internal tech. report, Aristotle University of Thessaloniki, Greece, Automation and Robotics Lab., Dept. of Electrical and Computer Eng.

7.4 Aksela, M. (2003): Comparison of classifier selection methods for improving committee performance. In Windeatt, T., Roli, F., editors, Proceedings of the 4th International Workshop on Multiple Classifier Systems, Lecture Notes in Computer Science, Guilford, UK. Springer-Verlag, 84–93

7.5 Back, T. (1996): Evolutionary algorithms in theory and practice: evolution strategies, evolution programming, genetic algorithms. Oxford University Press, Inc

7.6 Banfield, R. E, Hall, L. O., Bowyer, K. W., Kegelmeyer, W. P. (2003): A new ensemble diversity measure applied to thinning ensembles. In Windeatt, T., Roli, R. editors, *Proceedings of the 4th International Workshop on Multiple Classifier Systems*, Lecture Notes in Computer Science, Guilford, UK. Springer-Verlag, 306–316

7.7 Blake, C. L., Merz, C. J. (1998): UCI repository of machine learning databases, html://www.ics.uci.edu/ mlearn/mlrepository.html

7.8 Breiman, L. (2000): Randomizing output to increase prediction accuracy. Machine Learning, **40**, 229–242

7.9 Cho, S. B., Ahn, J. H. (2001): Speciated neural networks evolved with fitness sharing technique. In Proceedings of the 2001 Congress on Evolutionary Computation, **1**, 390–396

7.10 Cho, S. B., Ahn, J. H., Lee, S. I. (2001): Exploiting diversity of neural ensembles with speciated evolution. In Proceedings. IJCNN '01. International Joint Conference on Neural Networks, **2**, 808–813

7.11 Christensen, S. W. (2003): Ensemble construction via designed output distortion. In Windeatt, T., Roli, R., editors, *Proceedings of the 4th International Workshop on Multiple Classifier Systems*, Lecture Notes in Computer Science, Guilford, UK. Springer-Verlag, 286–295

7.12 Coello, C. A. C. (2001): A short tutorial on evolutionary multiobjective optimization. In Zitzler, E., Deb, K., Thiele, T., Coello, C. A. C., Corne, D., editors, First International Conference on Evolutionary Multi-Criterion Optimization, Lecture Notes in Computer Science No. 1993, Springer-Verlag, 21–40

7.13 Cunningham, P., Carney, J. (2000): Diversity versus quality in classification ensembles based on feature selection. In 11th European Conference on Machine Learning, Barcelona, Catalonia, Spain. Springer, Berlin, volume 1810, 109–116

7.14 Dietterich, T. G. (2000): Ensemble methods in machine learning. Lecture Notes in Computer Science No. 1857, Springer-Verlag

7.15 Freund, Y., Schapire, R. (1996): Experiments with a new boosting algorithm. In Proceedings of the Thirteen International Conference on Machine Learning, Morgan Kaufmann, 149–156

7.16 Fumela, G., Roli. R. (2002): Performance analysis and comparison of linear combiners for classifier fusion. In Caelli, T., Amin, A., Duin, R. P. W., Kamel, M. S., de Ridder, D., editors, Structural, Syntactic, and Statistical Pattern Recognition, Joint IAPR International Workshops SSPR 2002 and SPR 2002, Proceedings, volume 2396 of Lecture Notes in Computer Science, Windsor, Ontario, Canada. Springer, 424–432

7.17 Hashem, S. (1996): Effects of collinearity on combining neural networks. Connection Science: Specaill Issue on Combining Artificial Neural Networks, Ensemble Approaches, 8, 315–336

7.18 Haykin, S. (1999): Neural networks : a comprehensive foundation. Prentice-Hall

7.19 Jimenez, D., Walsh, N. (1998): Dynamically weighted ensemble neural networks for classification. In Proceedings of the 1998 International Joint Conference on Neural Networks, 753–756

7.20 Krogh, A., Vedelsby, J. (1995): Neural network ensembles, cross validation and active learning. In Tesauro, G., Touretzky, D. S., Len, T. K., editors, Advances in Neural Information Processing System, 7, MIT Press, 231–238

7.21 Kuncheva, L. I. (2003): Error bounds for aggressive and conservative adaboost. In Windeatt, T., Roli, F., editors, Proceedings of the 4th International Workshop on Multiple Classifier Systems, Guilford, UK, Lecture Notes in Computer Science, Springer-Verlag, 25–34

7.22 Lazarevic, A., Obradovic, Z. (2001): Effective pruning of neural network classifier ensembles. In Proceedings. International Joint Conference on Neural Networks, 2, 796–801

7.23 Kuncheva L. I., (2003): That elusive diversity in classifier ensembles. In Proc IbPRIA 2003, Mallorca, Spain, 2003, Lecture Notes in Computer Science, Springer-Verlag, 1126–1138

7.24 Lin, G., Yao, X., Macleod, I. (1996): Parallel genetic algorithm on pvm. In Proceedings of the International Conference on Parallel Algorithms, 1, 605–610

7.25 Liu, Y., Yao, X. (1997): Evolving modular neural networks which generalise well. In Proc. of 1997 IEEE International Conference on Evolutionary Computation, Indianapolis, USA, 605–610

7.26 Liu, Y., Yao, X. (1999): Ensemble learning via negative correlation. Neural Networks, 12, 1399–1404

7.27 Liu, Y., Yao, X., Higuchi, T. (2000): Evolutionary ensembles with negative correlation learning. IEEE Trans. Evolutionary Computation, 4, 380–387.

7.28 McKay, R., Abbass, H. A. (2001): Anti-correlation: a diversity promoting mechanism in ensemble learning. The Australian Journal of Intelligence Information Processing Systems, 7, 139–149

7.29 Navone, H. D., Verdes, P. F., Granitto, P. M., Ceccatto, H. A. (2000): Selecting diverse members of a neural network ensemble. In SBRN 2000, VI Brazilian Symposium on Neural Networks, Rio de Janeiro, Brazil.

7.30 Oliveira, L. S., Sabourin, R., Bortolozzi, F., Suen, C. Y. (2003): Feature selection for ensembles: a hierarchical multi-objective genetic algorithm approach. In 7th International Conference on Document Analysis and Recognition (ICDAR2003), IEEE Computer Society, Edinburgh-Scotland, 676–681

7.31 Optiz, D. W., Shavlik, J. W. (1996): Actively searching for an effective neural network ensemble. Connection Science. Special Issue on Combining Artificial Neural: Ensemble Approaches, 8, 337–354.

7.32 Oza, N. C. (2003): Boosting with averaged weight vectors. In Windeatt, T., Roli, F., editors, Proceedings of the 4th International Workshop on Multiple Classifier Systems, Guilford, UK, Lecture Notes in Computer Science, Springer-Verlag, 15–24

7.33 Pennock, D. M., Maynard-Reid II, P., Giles, C. L., Horvitz, E. (2000): A normative examination of ensemble learning algorithms. In Proc. 17th International Conf. on Machine Learning, Morgan Kaufmann, San Francisco, CA, 735–742

7.34 Perrone, M. P., Cooper, L. N. (1993): When networks disagree: ensemble methods for hybrid neural networks. In Mammone, R. J., editor, Neural Networks for Speech and Image Processing, Chapman-Hall, 126–142

7.35 Raviv, Y., Intrator, N. (1999): Variance reduction via noise and bias constraints. In Sharkey, A. J. C. editor, Combining artificial neural nets, 31. Springer-Verlag

7.36 Renner, R. S. (1999): Improving generalization of constructive neural networks using ensembles. Phd, The Florida State University

7.37 Rosen, B. E. (1996): Ensemble learning using decorrelated neural networks. Connection Science. Special Issue on Combining Artificial Neural: Ensemble Approaches, 8, 373–384

7.38 Schapire, R. E. (1990): The strength of weal learnability. Machine Learning, 5, 197–227

7.39 Schapire, R. E. (1999): A brief introduction to boosting. In Proceedings of the Sixteenth International Joint Conference on Artificial Intelligence, 1401–1406

7.40 Schwefel, H. P. (1981): Numerical optimization for computer models. John Willey, Chichester, U.K.

7.41 Sharkey, A. (1998): Combining artificial neural nets: ensemble and modular multi-net systems. Springer-Verlag, New York, Inc.

7.42 Sharkey, A. J. C. (1996): On combining artificial neural nets. Connection Science, 8, 299–313

7.43 Skalak, D. B. (1996): The sources of increased accuracy for two proposed boosting algorithms. In Proceedings of the AAAI-96 Integrating Multiple Learned Models Workshop.

7.44 Standish, P. K. (2002): Diversity evolution. In Standish, P. K., Bedau, M., Abbass, H. A. editors, Proceedings Artificial Life VIII, Cambridge MA: MIT Press, 131–137

7.45 Turner, K., Ghosh, J. (1996): Error correlation and error reduction in ensemble classifiers. Connection Science, 8, 385–403

7.46 Ueda, N., Nakano, R. (1996): Generalization error of ensemble estimators. In IEEE International Conference on Neural Networks, 1, 90–95

7.47 Van Veldhuizen, D. A., Lamont, G. B. (2000): Multiobjective evolutionary algorithms: analyzing the state-of-the-art. Evolutionary Computation, 8, 125–147

7.48 Wu, J. X., Zhou, Z. W., Chen, Z. Q. (2001): Ensemble of ga based selective neural network ensembles. In Proceedings of the 8th International Conference on Neural Information Processing, 3, 1477–1482

7.49 Bouckaert, R. (2002): Accuracy bounds for ensembles under 0-1 loss. Technical report, Mountain Information Technology Computer Science Department, University of Waikato, New Zealand.

7.50 Yao, X. (1999): Evolving artificial neural networks. Proceedings of the IEEE, 87, 1423–1447

7.51 Yao, X., Liu, Y.(1996): Ensemble structure of evolutionary artificial neural networks. In IEEE International Conference on Evolutionary Computation (ICEC'96), Nagoya, Japan, 659–664

7.52 Yao, X., Liu, Y. (1997): A new evolutionary system for evolving artificial neural networks. IEEE Transactions on Neural Networks, 8, 694–713

7.53 Yao, X., Liu, Y. (1998): Making use of population information in evolutionary artificial neural networks. IEEE Transactions on Systems, Man, and Cybernetics, Part B: Cybernetics, **28**, 417–425

7.54 Zenobi, G., Cunningham, P. (2001): Using diversity in preparing ensemble of classifiers based on different subsets to minimize generalization error. In Lecture Notes in Computer Science, **2167**

7.55 Zhou, Z. H., Wu, J., Tang, W. (2002): Ensembling neural networks: many could be better than all. Artificial Intelligence, **137**, 239–263

7.56 Zhou, Z. H., Wu, J. X., Jiang, Y., Chen, S. F. (2001): Genetic algorithm based selective neural network ensemble. In Proceedings of the Seventeenth International Joint Conference on Artificial Intelligence, 797–802

7.57 Zitzler, E. (1999): Evolutionary algorithms for multiobjective optimization: methods and applications. Phd thesis, Swiss Federal Institute of Technology (ETH) Zurich

8. Unsupervised Niche Clustering: Discovering an Unknown Number of Clusters in Noisy Data Sets

Olfa Nasraoui[1], Elizabeth Leon[1], and Raghu Krishnapuram[2]

[1] Department of Electrical and Computer Engineering
The University of Memphis
206 Engineering Science Bldg.
Memphis, TN 38152-3180
onasraou@memphis.edu

[2] IBM India Research Lab
Block 1, Indian Institute of Technology
Hauz Khas, New Delhi 110016, India
kraghura@in.ibm.com

Abstract: As a valuable unsupervised learning tool, clustering is crucial to many applications in pattern recognition, machine learning, and data mining. Evolutionary techniques have been used with success as global searchers in difficult problems, particularly in the optimization of non-differentiable functions. Hence, they can improve clustering. However, existing *evolutionary* clustering techniques suffer from one or more of the following shortcomings: **(i)** they are *not robust* in the presence of noise, **(ii)** they assume a *known* number of clusters, and **(iii)** the size of the search space *explodes exponentially* with the number of clusters, or with the number of data points. We present a *robust* clustering algorithm, called the *Unsupervised Niche Clustering algorithm (UNC)*, that overcomes all the above difficulties. UNC can successfully find dense areas (clusters) in feature space and determines the *number* of clusters *automatically*. The clustering problem is converted to a multimodal function optimization problem within the context of Genetic Niching. Robust cluster scale estimates are *dynamically* estimated using a hybrid learning scheme coupled with the genetic optimization of the cluster centers, to adapt to clusters of different sizes and noise contamination rates. Genetic optimization enables our approach to handle data with both numeric and qualitative attributes, and general *subjective, non metric, even non-differentiable* dissimilarity measures.

8.1 Introduction

Clustering [8.10, 8.15, 8.1] is an effective technique for data mining and exploratory data analysis that aims at classifying the unlabelled points in a data set into different groups or clusters, such that members of the same cluster are as similar as possible, while members of different clusters are as dissimilar as possible. Several approaches to clustering exist. For example, graph-theoretic and tree-based techniques are popular in the machine learning community; while most objective function-driven or prototype-based clustering methods such as the $K-$Means [8.29], its fuzzy counterpart, the Fuzzy C-Means (FCM) [8.37, 8.11, 8.3], and Gaussian Mixture modelling, have long been used in statistical pattern recognition. Recently, data mining has put even higher demands on clustering algorithms. They now must handle very large data sets, leading to many scalable clustering techniques. Examples are CLARANS [8.31] and BIRCH [8.39], which assume that clusters are hyper-spherical, similar in size, and span the whole data space. Other techniques include CURE [8.19] and the scalable K-Means and the scalable EM [8.4, 8.5]. Unfortunately, all the above techniques were not designed to handle large and unknown amounts of noise in the data. *Robust* clustering techniques have recently been proposed to handle noisy data. Another limitation of most clustering algorithms is that they assume that the number of clusters is known.

8.1.1 Unsupervised Clustering

When the number of clusters is not known, the situation is sometimes called *unsupervised clustering*. Traditionally, unsupervised clustering has relied on three different approaches. The first approach is to evaluate the validity of the partition generated by the clustering for several values of number of clusters, c, and then accept the partition with optimal validity. The second approach is to seek and remove one cluster at a time provided that the found cluster passes a validity test [8.23]. The problem with these approaches lies in the difficulty in designing validity measures that perform well on a variety of data sets encountered in practice. Also, most validity measures either assume a known underlying inlier or noise distribution or are very sensitive to noise, and hence are not appropriate for general robust clustering. The third approach consists of starting the clustering process with an over-specified number of clusters, and then merging similar clusters and eliminating spurious clusters until the correct number of clusters is left as in Compatible Cluster Merging [8.25].

8.1.2 Motivations for Evolutionary Clustering

Inspired by nature, Genetic Algorithms (GAs) [8.21] and evolutionary strategies [8.13] constitute a powerful set of global search techniques that have demonstrated good performance on a wide variety of problems. GAs search the solution space of a fitness function to be optimized using a simulated "Darwinian" evolution that favors survival of the fittest. Whereas traditional search techniques rely on characteristics of the objective function to be optimized (such as differentiability for gradients and Hessians, and sometimes linearity and continuity) to determine the next sampling point, GAs make no such assumption. Instead these points are determined based on stochastic sampling rules. This means that GAs can optimize fitness functions of many forms, subject to the minimal requirement that the function can map the population into a partially ordered set [8.17].

The Simple Genetic Algorithm (SGA) has been successfully used to search the solution space in clustering problems with a *fixed* number of clusters [8.20], and for robust clustering [8.33]. However, in practice, the number of clusters may not be known.

8.1.3 Why Genetic Niching?

There is a *symbiosis* between the way *niches* evolve in nature and the way data is partitioned into optimal *clusters*. In the Evolutionary Computation area, Genetic Niching (GN) techniques have emerged, largely inspired by nature, to solve the multimodal function optimization problem and to *counteract premature population convergence* by encouraging genetic diversity. In unsupervised learning, a wide variety of clustering techniques have been developed to *cluster heterogenous data into different components*.

We propose a novel approach to unsupervised robust clustering using Genetic Algorithms. We start by modifying our objective from searching the solution space for c clusters to searching this space for any one cluster. Accordingly, we need to optimize an appropriate objective function that simply measures the goodness of fit of a model to just part of the data. We formulate a density based fitness function that reaches a maximum at every good cluster center. This means that the fitness landscape will consist of multiple peaks with one at each cluster location. We use Deterministic Crowding (DC) [8.30] as the genetic niching optimization tool. To alleviate the problem of crossover interaction between distinct niches, we propose an improved restricted mating scheme which relies on an accurate and assumption-free estimate of the niche radii which are not restricted to be equal for all peaks. The resulting unsupervised clustering algorithm is called *Unsupervised Niche Clustering* (UNC). Since UNC is based on Genetic Optimization, it is much less prone to suboptimal solutions than traditional techniques. Moreover, because UNC uses robust weights, it is less sensitive to the presence of noise. Also, relying on Genetic optimization to search for the solution obviates the need to analytically derive the update equations for any prototypes. Thus, our approach is able to handle data with both numeric and qualitative attributes, and it can work with general subjective non metric dissimilarity measures. Handling such dissimilarities has traditionally been tackled by clustering relational data consisting of all pairwise dissimilarity values. In this case, the complexity of most algorithms, such as the Linkage type hierarchical clustering techniques can be prohibitively expensive (in the order of $\mathcal{O}\left(N^2 \log_2(N)\right)$).

The remainder of this chapter is organized as follows. In Section 8.2, we review niching methods. In Section 8.3, we survey some evolutionary clustering techniques. In Section 8.4, we describe our new approach to unsupervised clustering based on genetic niching. In Section 8.5, we present our synthetic experimental results. In Section 8.6, we apply our unsupervised clustering algorithm, UNC, to the problem of image segmentation. Finally, we present our conclusions in Section 8.7.

8.2 Genetic Niching

Traditional Genetic Algorithms have proved effective in exploring complicated fitness landscapes and converging populations of candidate solutions to a single global optimum. However, some optimization problems require the identification of global as well as local optima in a multimodal domain. As a result, several population diversity mechanisms have been proposed to delay or counteract the convergence of the population to a single solution by maintaining a diverse population of members throughout its search. As for the case of GAs, these diversity enhancing methods have turned to nature for ideas and analogies. An analogy to multimodal domains exists in nature in the form of "ecological niches" which are subspaces that can support different types of species or organisms. This has inspired the consideration of each peak in a multimodal domain as a niche in the framework of what has come to be called niche formation methods. In nature, the fertility of the niche as well as the efficiency of each organism at exploiting that fertility is what determines the number of organisms that can be contained in a niche. This principle is at the base of how a GA should maintain the population diversity of its members in a multimodal domain. Thus, the niches should be populated in proportion to their fitness relative to other peaks. This concept is known as niche proportionate population. This aim has been very difficult to realize because of the uncertainty in the niche location (peak *location* and niche *boundary*). The two most popular niche formation methods are sharing and crowding.

Sharing methods [8.21, 8.18] attempt to maintain a diverse population by reducing the fitness of individuals that have highly similar members within the population. This in turn discourages redundant solutions from overtaking the entire population, while rewarding individuals that uniquely exploit specific areas of the domain. Sharing methods rely strongly on correct estimates of the niche counts, i. e., the number of individuals in each niche. The niche counts themselves depend on a parameter, σ_{sh}, which ideally, should approximate the widths of the peaks. Deb and Goldberg [8.9] suggested ways of determining the appropriate value for σ_{sh} based on the expected number of peaks and the hypervolume of the entire domain space. Unfortunately, in many real applications, the number of peaks may not be known. Moreover, a single value of σ_{sh} may not be sufficient when peaks differ vastly in their widths. In addition to the above drawbacks, sharing adds an additional $\mathcal{O}(N_P^2)$ complexity per generation to compute all the pairwise distance calculations among all the members of a population of size N_P.

Crowding methods, proposed by De Jong in [8.24], try to form and maintain niches by replacing population members preferably with the most similar individuals. Unfortunately, stochastic "replacement errors" prevented this method from maintaining more than two peaks in a multi-modal fitness landscape. Mahfoud [8.30] proposed an improved crowding mechanism, called "deterministic crowding" (DC), which nearly eliminated replacement errors and proved more effective in maintaining multiple niches. DC is presented below:

Deterministic Crowding (DC)

Repeat for G generations {
  Repeat $\frac{N_P}{2}$ times {
    Select two parents p_1 and p_2 randomly without replacement;
    Cross them to produce children c_1 and c_2;
    Optionally apply mutation to produce children c_1' and c_2';
    IF $[d(p_1, c_1') + d(p_2, c_2')] \leq [d(p_1, c_2') + d(p_2, c_1')]$ THEN {
      IF $f(c_1') > f(p_1)$ THEN replace p_1 with c_1'
      IF $f(c_2') > f(p_2)$ THEN replace p_2 with c_2'
    }
    ELSE {
      IF $f(c_2') > f(p_1)$ THEN replace p_1 with c_2'
      IF $f(c_1') > f(p_2)$ THEN replace p_2 with c_1'
    }
  }
}

Unlike sharing methods, DC is free of any parameters relying on assumptions about the number of peaks or their widths. Also, unlike sharing, the expected distribution of the population at convergence is independent of fitness. Instead, the cardinality of each niche is expected to be proportional to the fraction of the population from the search space that falls within this niche, or in other words the prior probability of the niche. However, a dominated niche can, with another niche's assistance, cross to form members from a fitter (dominant) niche. This causes a migration of members from the dominated peaks to the dominant peaks that will only come to a halt when one of the dominated niches is depleted of its members. This crossover interaction problem can be critical for clustering because of the arbitrary density and position of clusters.

8.3 An Overview of Existing Evolutionary Clustering Techniques

Clustering techniques which use mathematical optimization methods or local search strategies are very sensitive to initialization, and can be trapped in local extrema. GAs perform a globalized search for solutions that can alleviate this problem. One of the earliest attempts at using a GA for clustering was made by Raghavan and Birchand [8.35]. In their approach, the GA was used to optimize the square error of clustering (similar to K Means's criterion). Each chromosome represented a possible partition of the entire data set consisting of N objects. Hence the chromosome consisted of N substrings, with each substring encoding one of c cluster labels. Obviously this encoding led to an explosion of the search space size as the data set got larger, and assumed a known number of clusters. The square error based fitness function also meant that the approach was sensitive to noise. Most importantly, the encoding scheme was poor, since n-point crossover frequently resulted in meaningless or lethal partitions. Bhuyan et al. [8.7] proposed an improved encoding that used a separator symbol ($*$) to separate the clusters, consisting of a string of data object labels, and Goldberg's permutation crossover [8.17] to yield valid offspring. However, the solution suffered from en explosion in permutation redundancy because of the arbitrary order of data labels. Babu and Murty [8.6] used the GA only to find good initial solutions, and then used K-Means for the final partition. This was the first hybrid clustering approach that obviously outperformed the use of either K-means or GA alone. However, it is not resistant to noise, and assumed a known number of clusters. Fogel and Simpson [8.14] use Evolutionary Programming to solve the fuzzy min-max cluster problem by directly encoding the centroids (instead of the entire partition) in the chromosome string. This was undoubtedly one of the first *efficient* encodings for the clustering problem, and it influenced most subsequent evolutionary clustering methods. Hall et al. [8.2, 8.20] proposed a genetically guided approach (GGA) to optimizing the reformulated Hard and Fuzzy C-Means (HCM and FCM) objective functions. The chromosome strings encode the c center vectors (of p individual features per vector) of the candidate solutions, and a standard GA evolves the population. However, GGA is *not robust* in the face of noise, and it *can not determine the number of clusters automatically*. Also, because all c cluster centers are encoded in each chromosome string, the size of the search space *explodes exponentially with the number of clusters*.

In [8.33], we proposed a robust estimator based on the LMedS that can simultaneously partition a given data set into c clusters, and estimate their parameters, for the case when the number of clusters, c is known a priori. In addition to its limitation of estimating the parameters of a single structure, the LMedS suffers from a major drawback in that it has a nonlinear, and non-differentiable objective function that is not amenable to mathematical or numerical optimization. For this reason, we proposed the integration of a genetic algorithm to the partitioning and estimation process, in order to search the solution space more efficiently. This resulted in a new approach to *robust genetic clustering* based on LMedS. However, the technique still assumed that the number of clusters was *known in advance*, and that the *noise contamination rate was* 50%. Also, because all c cluster centers are encoded in each chromosome string, the size of the search space explodes exponentially with the number of clusters as in the case of GGA.

Lee and Antonsson present an algorithm for unsupervised clustering using Evolutionary Strategies (ES) in [8.28]. In their approach, all cluster centroids are coded into a variable length chromosome, and only crossover is used to vary the number of clusters. However this coding scheme suffers from an exponential increase in the complexity of search with the number of clusters, and is also not robust to noise and outliers because its fitness measure is based on a classical sum of errors.

Rousseeuw's original robust K-medoid criterion [8.36] was optimized in [8.12] using a hybrid GA approach. Though more robust to noise, this approach assumes a known number of clusters. Also, because the cluster representatives are medians, it is most efficient when the rate of noise is exactly 50%, and cannot adapt to various noise contamination rates. It also does not have any provision for clusters of different sizes since it has no notion of scale, and the size of the search space explodes exponentially with the number of clusters.

In [8.34], we presented preliminary results of Unsupervised Niche Clustering (UNC). In this chapter, we present detailed derivations and discussions of the theoretical properties of UNC, as well as thorough evaluation experiments.

Table 8.1 compares some evolutionary clustering techniques including UNC with respect to important features such as robustness to noise, automatic determination of the number of clusters, $\cdots$, etc, showing that UNC has all the desired features lacking in other approaches. Note that a *partitional* approach relies on a sum of error type objective function that requires encoding of c cluster centers or a long c-ary partition string of length N. Partitional approaches also require the additional costly overhead of *re-partitioning the data points into c clusters with each fitness computation.* On the other hand, a *density* based approach directly optimizes each cluster density independently of other clusters, hence eliminating the need to partition data. Density is also a more sensible measure of cluster validity, and is naturally resistant to noise. Complexity is listed per generation. Also hybrid approaches tend to converge in fewer generations compared to purely evolutionary search methods, and are therefore faster.

8.4 New Approach to Unsupervised Robust Clustering using Genetic Niching

8.4.1 Representation

The solution space for possible cluster centers consists of n-dimensional prototype vectors. These are represented by concatenating the Gray codes of

Table 8.1. Comparison of UNC with other evolutionary clustering algorithms for data of size N, population of size N_P, and c clusters

Approach →	UNC	GGA [8.20]	Lee [8.28]	G-C-LMedS [8.33]	k-d-Median [8.12]
Search Method	GA	GA	ES	GA	GA
Robustness to noise	yes	no	no	yes	yes
Automatic Scale Estimation	yes	no	no	no	no
Complexity: $O()$	NN_P	CNN_P	CNN_P	CNN_P	$N_P CN \log(N)$
Hybrid	yes	no	no	no	yes
Does not require No. of Clusters	yes	no	yes	no	no
Handles ellipsoidal clusters	yes	no	no	no	no
Density (D)/ Partition (P)	D	P	P	P	P

the individual features for *a single* cluster center into a binary string. Paradoxically, this means that the search space is much smaller than the one corresponding to using the SGA to search for c cluster centers as in the case of the Fuzzy C-Means based GGA [8.2] and the Genetic C-LMedS [8.33]. If S is the search space size for our niching based unsupervised approach to clustering, then the search space size for the SGA based C-means clustering is S^c. As expected, the savings in the size of the search space translate into savings in the population size.

8.4.2 Fitness Function

Since in general, we identify dense areas of a feature space as clusters, we will define the fitness value, f_i, for a candidate center location, c_i, as the density of a hypothetical cluster at that location. For the case of an n-dimensional data set, $\mathcal{X}$, with N data points, $x_j, j = 1, \cdots, N$ the density of the ith cluster can be defined as follows

$$f_i = \frac{\sum_{j=1}^{N} w_{ij}}{\sigma_i^2}, \tag{8.1}$$

where

$$w_{ij} = \exp\left(-\frac{d_{ij}^2}{2\sigma_i^2}\right), \tag{8.2}$$

is a *robust* weight designed to be high for inliers (typical points falling within the boundaries of a cluster) and low for outliers (points that fall beyond the cluster boundaries), and d_{ij}^2 can be any distance measure, such as the Euclidean distance from data point $\mathbf{x}_j$ to cluster center $\mathbf{c}_i$,

$$d_{ij}^2 = \|\mathbf{x}_j - \mathbf{c}_i\|^2 = (\mathbf{x}_j - \mathbf{c}_i)^t (\mathbf{x}_j - \mathbf{c}_i). \tag{8.3}$$

σ_i^2 is a robust measure of scale (dispersion) for the ith cluster, which is crucial for determining its boundaries. In fact, the niche radius can be written as $K\sigma_i^2$, where K is close to $\chi_{n,.995}^2$ for spherical Gaussian clusters.

For the case of general multivariate hyper-ellipsoidal clusters, the fitness measure is evaluated as

$$f_i = \frac{\sum_{j=1}^{N} w_{ij}}{|\varSigma_i|^{\frac{1}{n}}}, \tag{8.4}$$

where

$$w_{ij} = \exp\left(-\frac{d_{Mij}^2}{2}\right), \tag{8.5}$$

In Eq. (8.4), $|.|$ is the determinant and $\varSigma_i$ is a dynamically estimated dispersion matrix, which is used to compute the Mahalanobis distance values that serve to compute the weights in Eq. (8.5). That is

$$d_{Mij}^2 = (\mathbf{x}_j - \mathbf{c}_i)^t \varSigma_i^{-1} (\mathbf{x}_j - \mathbf{c}_i). \tag{8.6}$$

It is important to observe that for the case of a diagonal dispersion matrix where all the diagonal entries are equal, both the fitness and robust weight in Eqs. (8.4) and (8.5) reduce to Eqs. (8.1) and (8.2) respectively. Even though Eq. (8.4) is more general than Eq. (8.1), when the dimensionality is high, there are several benefits in estimating scalar dispersion factors as opposed to dispersion matrices and their required inversions for each individual of the population. These benefits clearly come in the form of computational savings, but also in the form of better estimates because of the fewer number of parameters that have to be estimated. For general high dimensional data sets with arbitrary distributions, it may be preferable to assume a simpler model. Also, the simpler model with scalar dispersion allows more general dissimilarity measures to be used, even non-differentiable and non-metric ones. For this reason, we present the analysis and results of the approaches based on both the scalar and matrix dispersion models. For the case of $n = 2$, it can be seen that f_i is a measure of density since it measures the ratio of the robust cardinality, $\sum_{j=1}^{N} w_{ij}$, of the ith cluster or niche to its volume. For the case of $n > 2$, the niche volume is proportional to $|\varSigma_i|^{\frac{1}{2}}$, Hence, the fitness is the density of the projection of the data points on a 2-dimensional subspace, where the resulting dispersion is proportional to the geometrical mean of the dispersions along the individual dimensions of the data set. In this case, σ_i^2 and $|\varSigma_i|^{\frac{1}{n}}$ become analogous, i.e., they are both approximately proportional to the square of the projected niche radius for the ith cluster. In both cases, the normalization of the cardinality of the niche by its size gives preference to individuals located in denser niches.

8.4.3 Analogy between Density and Scale in Artificial Niches and Fertility and Barrier Constraints in Natural Niches

One way to dynamically estimate the niche sizes or scales is to compute the optimal scale values that will maximize the proposed density fitness functions above. This is because the scale that will maximize cluster or niche density corresponds to the optimal niche size that achieves the right balance between niche compactness (as measured by the within niche dispersion in the denominator) and niche count (as measured by the soft cardinality in the numerator). Minimizing the dispersion (denominator) *encourages proximity* of members of the same niche, hence promoting the establishment of *barriers* between distinct niches; while maximizing the soft cardinality in the numerator avoids the degenerate case that would result from optimizing compactness alone, and that can result in a niche accepting no more than a single member. Hence it encourages the *inclusion of as many members as possible* to populate the niche as long as enough resources are available. Hence, optimizing the above density fitness functions is consistent with the way that ecological niches are formed. They are areas that attract relatively higher members of the population compared to their surroundings.

8.4.4 A Baldwin Effect for Scale Estimation

The scale parameter or niche radius that maximizes the fitness value in Eq. (8.1), for the ith cluster can be found by setting $\frac{\partial f_i}{\partial \sigma_i^2} = 0$ to obtain

$$\sigma_i^2 = \frac{1}{2} \frac{\sum_{j=1}^{N} w_{ij} d_{ij}^2}{\sum_{j=1}^{N} w_{ij}}. \tag{8.7}$$

The centers at which density is maximized are unbiased estimates of the cluster centroids (this can easily be verified by mathematical optimization of the fitness criterion). However, the scale values in Eq. (8.7) tend to be underestimated as seen for the simple case of a single cluster, in Fig. 8.1(a).

We compensate for this underestimation by adjusting the bias for the scale so that, if the statistical measure of variance, $\sigma_s^2 = E(\mathbf{x}_j - \mu_i)^2$, is considered as the reference, the bias would be zero (i.e., the estimate becomes unbiased). The adjustment of bias of σ_i^2 with respect to σ_s^2, denoted as $B_{\sigma_s^2}(\sigma_i^2)$ below, proceeds by adjusting the scale estimates σ_i^2 by a multiplying factor, α that will make its bias zero, as follows:

$$B_{\sigma_s^2}(\alpha\sigma_i^2) = E(\alpha\sigma_i^2) - \sigma_s^2 = E\left(\frac{\alpha}{2} \frac{\sum_{j=1}^{N} w_{ij} d_{ij}^2}{\sum_{j=1}^{N} w_{ij}}\right) - \sigma_s^2 = 0.$$

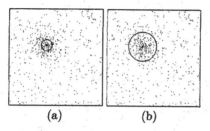

$$(a) \qquad\qquad (b)$$

Fig. 8.1. Effect of bias adjustment on scale estimates:(a) results with a biased estimate (b) results with the compensated unbiased estimate

Fixing the weights w_{ij} as was done before deriving the optimal scale values in each iteration, we obtain

$$B_{\sigma_s^2}(\alpha\sigma_i^2) = \frac{\alpha}{2}E(\sum_{j=1}^{N} \frac{w_{ij}}{\sum_{k=1}^{N} w_{ik}} d_{ij}^2) - \sigma_s^2 = \frac{\alpha}{2}\sum_{j=1}^{N} \frac{w_{ij}}{\sum_{k=1}^{N} w_{ik}} E(d_{ij}^2) - \sigma_s^2 = 0$$

$$B_{\sigma_s^2}(\alpha\sigma_i^2) = \frac{\alpha}{2}E(d_{ij}^2) - \sigma_s^2 = \frac{\alpha}{2}E(\mathbf{x}_j - \mu_i)^2 - \sigma_s^2 = 0.$$

This reduces to

$$B_{\sigma_s^2}(\alpha\sigma_i^2) = \frac{\alpha}{2}E(\mathbf{x}_j - \mu_i)^2 - \sigma_s^2 = \frac{\alpha}{2}\sigma_s^2 - \sigma_s^2 = 0$$

which is satisfied iff $\alpha = 2$.

Hence, the unbiased scale parameter or niche radius $(\alpha\sigma_i^2)$ for the ith cluster becomes

$$\sigma_i^2 = \frac{\sum_{j=1}^{N} w_{ij}d_{ij}^2}{\sum_{j=1}^{N} w_{ij}}. \tag{8.8}$$

Therefore, the σ_i^2 will be updated using an iterative hill-climbing procedure, using the previous values of σ_i^2 to compute the weights w_{ij} in Eq. (8.2). Note that this approximation is reasonable because by virtue of the replacement scheme in DC, the centers for an individual are not expected to change drastically from one generation to the next. Even though the exponential weights w_{ij} decrease with distance, points from other clusters can still exert an adverse influence on a given cluster's parameter estimates in the early generations, particularly when the data set contains a large number of clusters. For this reason, before updating the scale parameters, the weights will be mapped to a binary range. The resulting hard rejection will gradually allow only the core members of the cluster to be involved in the estimation of its parameters. The binarization of the weights is done as follows

$$w_{ij} = \begin{cases} 1 \text{ if } w_{ij} > T_w \\ 0 \text{ otherwise} \end{cases} \tag{8.9}$$

where T_w is a suitable threshold value.

The dispersion matrix that maximizes the fitness value in Eq. (8.4), for the ith cluster can be found by setting $\frac{\partial f_i}{\partial \Sigma_i^{-1}} = 0$ and solving for Σ_i to obtain

$$\Sigma_i = \frac{\sum_{j=1}^{N} w_{ij}(x_j - c_i)(x_j - c_i)^T}{\sum_{j=1}^{N} w_{ij}}. \tag{8.10}$$

Therefore, when the fitness measure depends on a dispersion matrix, the Σ_i will be updated using an iterative hill-climbing procedure, using the previous values of Σ_i to compute the weights w_{ij} in Eq. (8.5). As in the case of scalar dispersions, the weights are also binarized to counteract the initial influence of points from different niches.

The proposed approach to estimating the niche sizes makes our approach a hybrid genetic optimization technique. Unlike *Lamarckian learning* [8.38], our approach to estimate the scale *does not disrupt the genotype* of the candidate solutions. However, it improves *individual learning* in the evolutionary process by dynamically modifying the *fitness landscape* (better scale estimates will result in more accurate cluster/niche density estimates) in a way that will make it easier to maintain diversity and to converge closer to the niche peaks. This can be seen as introducing a *Baldwin effect* [8.22] to the evolutionary process.

8.4.5 Mating Restriction

We impose some restrictions on the DC selection and replacement procedure described in Section 8.2 to prevent the production of lethal offspring or extinction of certain dominated niches. Let s_i denote a measure of the current scale estimate of individual P_i, where $s_i = \sigma_i^2$ if the scalar dispersion model is used and $s_i = |\Sigma_i|^{\frac{1}{n}}$ if the matrix dispersion model is used. Two population members P_i and P_j, with corresponding centers c_i and c_j, are considered to be coming from different niches if neither one of them is within the other member's niche. Then P_i and P_j are considered to be from distinct niches if P_j is not within P_i's niche, or $dist(P_i, P_j) > K s_i$; and P_i is not within P_j's niche, or $dist(P_i, P_j) > K s_j$, where $dist(P_i, P_j) = \|c_i - c_j\|^2$. These two conditions are equivalent to $dist(P_i, P_j) > K \max(s_i, s_j)$. If parents from different niches are forbidden from mating regardless of their fitness values, then mediocre individuals located in non-niche areas of the search space will not be allowed to improve their genotype. As a result, these undesirable individuals which can result from the random initialization process, particularly when the data set contains outliers, will remain in the population until convergence. Therefore, mating restriction should be relaxed by allowing unconditional mating between two individuals if at least one of them has low fitness. The resulting mating restriction rule for individuals P_i and P_j can be summarized as

$$\text{IF } (dist\,(P_i, P_j) > K \max{(s_i, s_j)} \text{ and } f_i > f_{min} \text{ and } f_j > f_{min}) \atop \text{THEN Restrict Mating.} \qquad (8.11)$$

8.4.6 Scale Inheritance

When mating takes place, each child inherits the scale parameter, σ_i^2, or dispersion matrix, Σ_i, of the closest parent as its initial scale. This inheritance is possible because in DC, children are most likely to replace similar parents. This makes it possible to integrate a hill-climbing step that encompasses all generations for the scale parameter updates. Our *one-step* local Baldwin Effect learning approach, enabled by a cumulative *memory* of past generations, is more efficient than traditional hybridizations of GA and hill-climbing procedures because the latter iteratively update the parameters within each generation. However, these updates cannot be used as starting points for the next generation due to arbitrary changes in the population distribution between generations, particularly in the early generations (no memory).

The *full* or *from-scratch* learning of existing hybrid approaches, while making evolution faster (i.e., fewer generations are needed to converge), suffers from an added heavy computational cost in *each* generation for *each* individual. Our approach has the advantage of not requiring more than a single step of updating the scales, and hence adapting the environment (fitness) to accelerate evolution. Following their inheritance by both children, the scale parameters are updated using Eq. (8.8). Similarly, the parents' scale parameters are updated so that both children and parents' scale parameters would have implicitly undergone the same number of updates starting from the initial generation. This is essential to make a fair comparison of their fitness for the replacement decision in DC. When mating is not allowed, the parents' scale parameters are updated, their fitness values are recomputed, and they remain in the population for the next generation.

8.4.7 Crossover and Mutation

When mating between two individuals is allowed, one-point crossover is performed independently on each of the string sections representing the individual feature dimensions of the candidate cluster centers. This leads to n independent crossovers per offspring, each with a crossover probability P_c. After mutation, each bit in an offspring individual's chromosome string can be inverted with a small mutation probability P_m.

8.4.8 Incorporating Constraints Against Degenerate and Spurious Solutions

When an individual in the population is initialized at a remote location in a data set, its scale parameter will either increase excessively so that more inliers are included within its boundaries, or conversely shrink towards zero to achieve minimum size. These degenerate solutions result in high or invalid fitness respectively. We can limit this behavior by ensuring that the scale parameter does not exceed a theoretical upper bound on σ_i^2. For example, if the data set consists of only one Gaussian cluster, with center $\mathbf{c}$ and scale parameter σ^2, then we have $\max_{i=1}^{N} dist^2(\mathbf{x}_i, \mathbf{c}) = \frac{\max_{i,j=1}^{N} dist^2(\mathbf{x}_i,\mathbf{x}_j)}{4}$. Since we can approximate $\frac{\max_{i=1}^{N} dist^2(\mathbf{x}_i,\mathbf{c})}{\sigma^2}$ with $\chi_{n,0.995}^2$, it follows that

$$\frac{\max_{i,j=1}^{N} dist^2(\mathbf{x}_i, \mathbf{x}_j)}{(2\sigma)^2} \approx \chi_{n,0.995}^2. \tag{8.12}$$

Based on Eq. (8.12), an upper bound for σ_i^2 can be computed easily using a rough estimate of the diameter of the entire data set or equivalently the maximum distance between any two data points in $\mathcal{X}$. This results in

$$\sigma_{max}^2 = \frac{\sum_{p=1}^{n} \left(\max_{j=1}^{N} x_{jp} - \min_{j=1}^{N} x_{jp}\right)^2}{4\chi_{n,.995}^2}. \tag{8.13}$$

Therefore, σ_i^2 will not be updated if the updated value will exceed σ_{max}^2 or falls below $\sigma_{min}^2 = \frac{\sigma_{max}^2}{100}$. In the case of ellipsoidal clusters, we have noticed that poor individuals will have their scale matrix, Σ, either shrink or expand in only one dimension to yield extremely elongated clusters. While $s = |\Sigma|^{\frac{1}{n}}$ would shrink toward zero in the first case, it tends to remain within normal bounds in the second case. Hence, we need to rely on $\max_{k=1}^{n} \Sigma_{kk}$ to detect the case of an expansion situation. Hence, we will suppress the fitness of an individual that satisfies $\max_{k=1}^{n} \Sigma_{kk} > \sigma_{max}^2$ In addition to degenerate solutions with an invalid scale parameter, the population may contain spurious clusters with very few points. In general, they will form their own niches in the sparse areas of the feature space, and remain in the population owing to the diversity-maintaining nature of DC. For these individuals, the scale may also shrink toward zero. In order to eliminate these individuals, we start by forcing the fitness of an individual to be zero when its variance falls below the bound, $\sigma_{min}^2 = \frac{\sigma_{max}^2}{100}$, or when its robust cardinality, $\sum_{j=1}^{N} w_{ij}$, falls below an application dependent lower limit, N_{min} (lowest acceptable cluster cardinality). Therefore we modify the fitness function to become

$$f_i = \begin{cases} \frac{\sum_{j=1}^{N} w_{ij}}{\sigma_i^2} & \text{if } \sigma_{min}^2 < s_i < \sigma_{max}^2 \text{ and } \sum_{j=1}^{N} w_{ij} \geq N_{min} \\ 0 & \text{otherwise.} \end{cases} \tag{8.14}$$

Consequently, an individual with zero fitness can never survive into the next generation because in DC, a better fit child will always replace its parent. In addition, we prevent the child that is most similar to a zero fitness parent from replacing this parent to limit the propagation of mediocrity into future generations. This is done by modifying the DC replacement rule whenever a zero fitness or invalid parent is selected to contribute in a crossover operation.

In this case, regardless of whether crossover takes place, the invalid parent is replaced by the other parent. However, if the latter is also invalid, then it is replaced by a valid member that is randomly selected from the population. Note that we allow zero fitness parents to reproduce when mating is not restricted because by mating with individuals from different niches, it is possible to generate individuals in new unexplored areas of the feature space, particularly in the early generations. In other words, they are allowed to mate to enhance the exploration capabilities of the genetic optimization.

8.4.9 Selecting the Initial Population

The most rudimentary approach to initialization would be to select bit values randomly to build the chromosome strings of the initial individuals. However, this does not exploit the special characteristics of the clustering problem. Since the optimal solution consists of a cluster center, it is natural to select the initial centers randomly from the set of feature vectors. This results in a population of N_P individuals, P_i, $i = 1, \cdots, N_P$. The initial scale parameters are initialized using a fraction of the upper bound in Eq. (8.13), $\frac{\sigma_{max}^2}{K_\sigma}$. When dispersion matrices are used, they are all initialized to $\frac{\sigma_{max}^2}{K_\sigma} I_n$, where I_n is the $n \times n$ identity matrix. $K_\sigma = 10$ was used in all our experiments.

8.4.10 Extracting Cluster Centers From the Final Population

After convergence of the population, almost all individuals converge to the niche peaks or cluster centers. At this point, we extract the best individual from each good niche to obtain the set of final cluster centers, $\mathcal{C}$. We start by sorting the members of the final population in descending order of their fitness values, so that the better fit individuals are extracted before the lesser fit ones. Then, the best individual is extracted as the first cluster center. Subsequently, each good member of the sorted population is considered as a candidate for inclusion in the list of final cluster centers. For this purpose, the candidate center is compared to all the cluster centers that have already been extracted. The candidate center, P_i is considered to be similar to one of the previously extracted centers, P_k, if P_k is within P_i's niche and P_i is within P_k's niche. If the fitness measures depend on scalar dispersion as in Eq. (8.1), then this means that $dist(P_i, P_k) \le K\sigma_i^2$ and $dist(P_i, P_k) \le K\sigma_k^2$. These two conditions are equivalent to $dist(P_i, P_k) \le K \min \left(\sigma_i^2, \sigma_k^2 \right)$. In other words, the following rule can be used to decide wether two individuals are form different niches

$$\text{IF } \left(dist(P_i, P_k) > K \min \left(\sigma_i^2, \sigma_k^2 \right) \right) \text{ THEN} \atop P_i \text{ and } P_k \text{ are from different niches.} \tag{8.15}$$

If the fitness measures depend on dispersion matrices as in Eq. (8.4), then the following equivalent rule can easily be derived to decide whether two individuals are form different niches (using the Mahalanobis distance measures d^2_{Mij} and d^2_{Mji} using P_i and P_j as reference niche peaks, respectively)

$$\text{IF } \left(\max\left(d^2_{Mij}, d^2_{Mji}\right) > K\right) \text{ THEN} \qquad (8.16)$$
$$P_i \text{ and } P_k \text{ are from different niches.}$$

When a candidate cluster is deemed to be similar to one of the extracted clusters, it is discarded and its comparison to the remaining extracted clusters is abandoned.

The extraction procedure is presented below:

Final Cluster Center Extraction

Sort individuals P_i in descending order of their fitness values to obtain $P_{(i)}$, $i = 1, \cdots, N_P$, such that $f_{(1)} \geq f_{(2)} \geq \cdots \geq f_{(N_P)}$;
Initialize set of cluster centers $\mathcal{C} = \emptyset$;
FOR $i = 1$ **TO** N_P **DO** {
    IF $f_{(i)} > f_{min\_extract}$ AND P_i and P_k are from different niches
    $\forall P_{(k)} \in \mathcal{C}$ THEN
    {
        $\mathcal{C} \leftarrow \mathcal{C} \cup P_{(i)}$;
    }
}

8.4.11 Refinement of the Extracted Prototypes

It is recommended that a local search be performed in the neighborhood of each solution provided by Genetic Optimization (GO) to increase accuracy. Fortunately, initialization becomes no longer an issue for local search methods, because GO is expected to yield a solution that is very close to the global optimum. To make the local refinement of the parameters of each cluster independent of other clusters, the data set is partitioned into c clusters before performing the local search, such that each feature vector is assigned to the closest prototype. Subsequently, the ith cluster is given by

$$\mathcal{X}_i = \left\{ \mathbf{x}_{(k)} \in \mathcal{X} \mid d^2_{ik} < d^2_{jk} \; \forall j \neq i \right\}, \text{ for } 1 \leq i \leq c. \text{ In } [8.32], \text{ we presented a}$$

new iterative robust estimator, called the Maximal Density Estimator (MDE) which can estimate the center and scale parameters accurately and efficiently (with linear complexity). MDE uses an alternating optimization of the centers using

$$\mathbf{c}_i = \frac{\sum_{\mathbf{x}_{(j)} \in \mathcal{X}_i} w_{ij} \mathbf{x}_j}{\sum_{\mathbf{x}_{(j)} \in \mathcal{X}_i} w_{ij}}, \qquad (8.17)$$

and the following update equation for the scale parameters

$$\sigma_i^2 = \frac{\sum_{\mathbf{x}_{(j)} \in \mathcal{X}_i} w_{ij} d^4_{ij}}{3 \sum_{\mathbf{x}_{(j)} \in \mathcal{X}_i} w_{ij} d^2_{ij}}. \qquad (8.18)$$

We have noticed that final refinement using MDE yields center and scale estimates that are significantly more accurate than the pure evolutionary search with UNC. On the other hand, UNC can be considered to provide a reliable initialization to MDE which is based on a local iterative search.

8.4.12 The Unsupervised Niche Clustering Algorithm (UNC)

The Unsupervised Niche Clustering Algorithm

Compute scale upper bound, σ_{max}^2 using Eq. (8.13);
Initialize population with N_P individuals, P_i, selected randomly from the data set;
Initialize scale parameters or dispersion matrices for $i = 1$ TO N_P:

$\sigma_i^2 = \frac{\sigma_{max}^2}{K_\sigma}$, or $\frac{\sigma_{max}^2}{K_\sigma} I_n$ respectively;

Update scale values, σ_i^2 or dispersion matrices Σ_i, using Eq. (8.8) or Eq. (8.10) respectively
Repeat for G generations {
  Repeat $\frac{N_P}{2}$ times {
    Select two parents p_1 and p_2 randomly without replacement;
    Decide whether Mating Restriction is to be imposed using Eq. (8.11);
    IF Mating NOT Restricted THEN {
      Cross them to produce children c_1 and c_2;
      Optionally apply mutation to produce children c_1' and c_2';
      Update scale values, σ_i^2 or dispersion matrices Σ_i, using Eq. (8.8) or Eq.
(8.10)
      respectively, for parents and children
      Compute fitness values, f_i, using Eq. (8.14), for parents and children
      IF $f_i = 0$ for any of the parents or children THEN
        Replace zero-fitness individual with another individual
          randomly selected from population
      Use Deterministic Crowding as a replacement strategy;
    }
    ELSE {
      Update scale values, σ_i^2 or dispersion matrices Σ_i, using Eq. (8.8) or Eq.
(8.10)
      respectively, for restricted parents
      Compute fitness values, f_i, using Eq. (8.14), for restricted parents
    }
  }
}
Extract final cluster centers and scales from the final population;
Refine final cluster centers and scales locally using Eq. (8.17) and Eq. (8.18)

8.4.13 Computational Complexity

In each generation, the most extensive computational requirement for UNC consists of computing the distance values involving N data points, fitness, and scale for each of the N_P individuals in the population, and $\frac{N_P}{2}$ inter-niche distances, resulting in $\mathcal{O}(N_P \times N)$ computations. The extraction step (done only once) requires sorting the fitnesses and computing the inter-niche distances, resulting in $\mathcal{O}\left(N_P\left(\log N_P + N_P\right)\right)$. From our experience, the population size in UNC, N_P, tends to be a very small fraction of the size of the data, N. Hence the complexity is approximately linear $\left(\mathcal{O}\left(N_P \times N\right)\right)$ with respect to the number of points. The number of iterations to converge is generally small (within 30 generations) due to the hybrid optimization, which makes it competitive even with respect to well known iterative methods such as the K-Means.

8.5 Simulation Results on Synthetic Examples

The examples used in this section illustrate the performance of UNC on spatial data in 2D, hence they are easier to inspect visually.

8.5.1 Detailed Phases of Cluster Evolution

Fig. 8.2 shows the evolution of the population (denoted by square symbols) using UNC for a noisy data set with 5 clusters. The initial population is chosen randomly from the set of feature vectors. This explains the higher concentration of solutions in the densest areas, which converge toward the correct centers in subsequent generations. The cluster center coordinates found using UNC, and refined using MDE, are shown in Figs. 8.2(e) and 8.2(f) respectively. The leveled contours correspond to boundaries including increasing quantiles ($\alpha = 0.25, 0.5, 0.75, 0.99$) of each cluster as derived from the normalized distance values $\frac{d_{ij}^2}{\sigma_i^2} = \chi_{2,\alpha}^2$. Hence, they indicate the accuracy and robustness of UNC's cluster scale estimates σ_i^2.

8.5.2 Sensitivity to GA Parameters, Noise, and Effect of Final Refinement

Our simulation results using various population sizes showed similar results as long as the population size exceeded 50 individuals. Convergence of the population occurred relatively fast, after about 30 generations. Hence, we show the effect of the only GA parameters that seemed to affect the results: crossover probability (P_c), and mutation probability (P_m). With each parameter combination, we performed 100 runs and show the averages of the following clustering quality measures: (i) Number of final extracted clusters, (ii) Average normalized centroid error. To compute the error measure in (ii), we pair each discovered cluster with the closest true cluster center, compute the Euclidean distance between them, normalize this distance by the maximum possible distance in a 256×256 image (i.e., $\sqrt{256^2 + 256^2}$), and finally average these pairwise distances over all discovered clusters. The

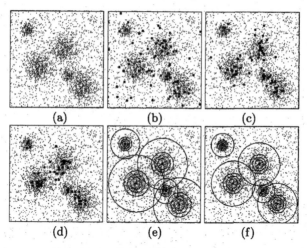

Fig. 8.2. Evolution of the population using UNQ: (a) original data set (b) initial population, (c) population after 10 generations,(d) population after 30 generations, (e) extracted centers and scales, (f) final centers and scales after refinement

GA parameters throughout all the experiments in this chapter were: population size = 80, number of generations = 30, $K = \chi^2_{2,.999} = 13.8$, $f_{min} = 0.6$, $f_{min\_extract} = 0.25$, and $T_w = 0.3$. We start by studying the effect of crossover and mutation as well as refinement on the clean and noisy data sets depicted in Fig. 8.3. The GA operator parameters were varied between 0.5 and 1 for P_c and between 0.001 and 0.2 for P_m. Figs. 8.4(a) and 8.5(a) show that for a wide range of crossover and mutation rates, UNC can discover the correct number of clusters in the vast majority of the experiments. Figs. 8.4(b) and 8.5(b) show that the accuracy of the center estimates, *without any final local refinement*, is within 1.5% of the range of the images, and that this accuracy improves with an increased crossover rate, while being stable for a wide range of mutation rates. The former is due to the fact that crossover encourages local exploration and recombination of good solutions to improve converged solutions. While the latter is due to Deterministic Crowding's replacement strategy which prevents lethal children resulting from mutation, from replacing their parents. The top two curves in each of Figs. 8.4(c) and 8.5(c) show that UNC is quite robust to noise, yielding comparable unrefined center estimate accuracies for both clean and noisy versions of the data sets, while the bottom two curves confirm that final local refinement dramatically improves this accuracy, as expected. Finally Fig. 8.6 shows how each iteration of the final local refinement improves the center estimates, hence reducing the bias, while also reducing the variance of the estimates, hence improving consistency across a whole range of experiments.

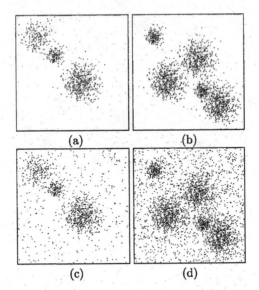

Fig. 8.3. Data sets with 3 and 5 clusters: (a-b) clean data sets (c-d) noisy data sets

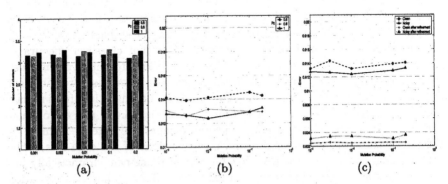

Fig. 8.4. Results of UNC for 3 clusters with noise, averaged over 100 runs:(a) average number of cluster for different crossover and mutation probabilities (b) center error, (c) center error after refinement ($P_c = 1$)

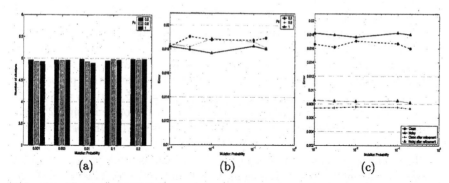

(a)   (b)   (c)

Fig. 8.5. Results of UNC for 5 clusters with noise, averaged over 100 runs:(a) average number of cluster for different crossover and mutation probabilities (b) Center Error, (c) center error after refinement ($P_c = 1$)

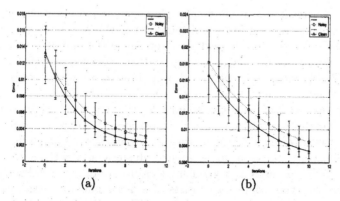

(a)   (b)

Fig. 8.6. Effect of final refinement (averaged over 100 runs with $P_c = 1, P_m = 0.001$) on average and standard deviation of center error for data sets with:(a) 3 clusters (b) 5 clusters

8.5.3 Simulation Results for Data with Varying Number of Clusters, Cluster Sizes, Densities, and Noise Contamination Rates

To illustrate UNC's performance under different conditions related to cluster size, density, noise contamination, orientation, and number of clusters, we present its results on the nine noisy data sets shown on Figs. 8.9(a) and 8.10(a) (first columns). The GA parameters were: population size $= 80$, number of generations $= 30$, crossover probability, $P_c = 1$, mutation probability, $P_m = 10^{-3}$, $K = \chi^2_{2,.999}$, $f_{min} = 0.6$, $f_{min\_extract} = 0.25$, and $T_w = 0.3$. The refinement of the centers and scale parameters is performed using 2 iterations of MDE for spherical clusters and 4 iterations of MMDE for ellipsoidal clusters.

The final centers computed using the Euclidean distance under the scalar dispersion model by UNC, K-Means [8.29] and the Possibilistic C-Means (PCM) [8.26, 8.27] are displayed in bold in Figs. 8.9(b) (second column), 8.9(c) (third column), and 8.9(d) (fourth column) respectively. Except for K-means which is not robust, and involves no scale estimation, the outermost circular contour around each cluster center depicts the normalized distances, $\frac{d_{ij}^2}{\sigma_i^2}$, corresponding to $\chi^2_{0.999}$, and hence encloses 99.9% of the data estimated to be inliers in each cluster. For PCM, σ_i^2 is considered to be the cluster scale parameter η_i which is initialized using the fuzzy average distance. For UNC, the levelled contours correspond to boundaries including increasing quantiles ($\alpha = 0.25, 0.5, 0.75, 0.99$) of each cluster as derived from the normalized distance values $\frac{d_{ij}^2}{\sigma_i^2} = \chi^2_{2,\alpha}$. Hence, they indicate accuracy and robustness of UNC's cluster scale estimates σ_i^2. Therefore, it reflects *robustness*. Note that unlike K-Means and PCM which are *provided with the correct number of clusters beforehand*, UNC succeeds in determining this number *automatically*. K-Means centers were initialized using randomly selected seeds from the data set, and PCM centers were initialized using randomly selected seeds from the data set which are then refined using 5 iterations of the Fuzzy C-Means. Both K-Means and PCM are sensitive to initialization and local minima, while K-Means suffers from a blatant lack of robustness. PCM suffers from another problem, that of often yielding *identical* clusters while missing other clusters. Fig. 8.7 illustrates the case where *no good* cluster exists in the data set. Note how only UNC succeeds in recognizing this situation. Fig. 8.8 illustrates the case where the whole data set consists of a dense uniform distribution, corresponding to a single cluster, and how UNC and PCM are able to detect a single cluster, while K-Means would naively partition the data into as many clusters as pre-specified.

Figs. 8.10(b) (second column) illustrate UNC's performance, using scale dispersion *matrices*, under different conditions related to *multivariate* cluster size, orientation, density, noise contamination, and number of clusters. For comparison purposes, Figs. 8.10(c) and (d) (third and fourth columns) show the results of K-Means and the Possibilistic C-Means (PCM) clustering algorithms, using the Gustafson-Kessel (GK) distance measure [8.16], on the same data sets. Note that unlike UNC which *determines the number of clusters automatically*, K-Means and PCM were *provided* with the correct number of clusters in advance. In these figures, the outermost ellipsoidal contour around each cluster center depicts the Mahalanobis distance, $d^2_{Mij} = \chi^2_{2,0.999}$; and the inner contours enclose increasing quantiles of each cluster as in the spherical case. Therefore, they reflect the accuracy of the final scale estimates.

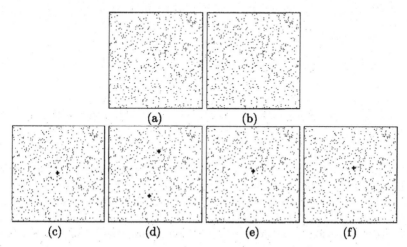

Fig. 8.7. Data set with no cluster: (a) original data set (b) results of UNC, (c) results of K-means with $C = 1$, (d) results of K-means with $C = 2$, (e) results of PCM with $C = 1$, (f) results of PCM with $C = 2$

8.6 Application to Image Segmentation

Image segmentation is a typical Multimedia data mining task, where different regions of an image are to be identified, for different applications ranging from image summarization, annotation, object recognition, motion analysis, to Content Based Image Retrieval. We use UNC to cluster the feature data extracted from the images shown on Fig. 8.11(a). Six features (3 color and 3 texture) are extracted from the original images in the COREL Image Database [?]. The Euclidean distance with scalar dispersion model is used. The GA parameters were: population size = 80, number of generations = 40, $P_c = .9$, $P_m = 5 \times 10^{-6}$, $K = \chi^2_{6,.995}$, $f_{min} = 1.0$, $f_{min\_extract} = 1.0$, and $T_w = 0.4$. Figs. 8.11(b) (second column) show the segmented images with a different color assigned to each cluster. The white pixels correspond to noise (robust weight $< w_{min} = \exp\left(\frac{-\chi^2_{6,.995}}{2\sigma_i}\right)$). These are pixels that are

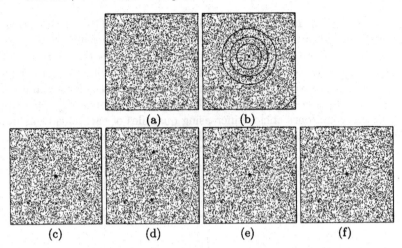

Fig. 8.8. Data set with uniform distribution:(a) original data set (b) results of UNC, (c) results of K-means with $C = 1$, (d) results of K-means with $C = 2$, (e) results of PCM with $C = 1$, (f) results of PCM with $C = 2$

either outliers or simply not at the core of a cluster. Also, often, the pixels lying on the boundaries between different regions end up being classified as noise (since they have low weight in either cluster), due to sampling and to the types of texture features which are region based. This is a natural side-effect of region based features, and can be improved using a variety of post-processing techniques.

Identifying noise or non-core pixels allows the delineation of cleaner (more homogenous) regions, and hence makes the computation of region validation measures more reliable and accurate. From a theoretical standpoint, noise identification is paramount to Content Based Image Retrieval and Object Recognition Systems, since this allows the computation of cleaner and more accurate similarity measures, and indirectly results in more accurate scale measures, which in turn result in more accurate confidence intervals, and more robust statistical tests.

Figs. 8.11(c) (third column) show the boundary images, where the black pixels correspond to pixels on the boundary between distinct regions. This illustrates another (dual) by-product of the segmentation process which is edge or boundary detection. The experiments are repeated with the matrix dispersion model for scale and fitness, and the segmentation and boundary results are shown in Figs. 8.12(b) and (c) respectively, using $T_w = 0.5$. Clearly, the results obtained using Euclidean distance and a scalar dispersion show more region fragmentation (more clusters per region) than their corresponding Mahalanobis distance based and dispersion matrix segmentations. This is because even naturally elongated clusters in feature space tend to be split

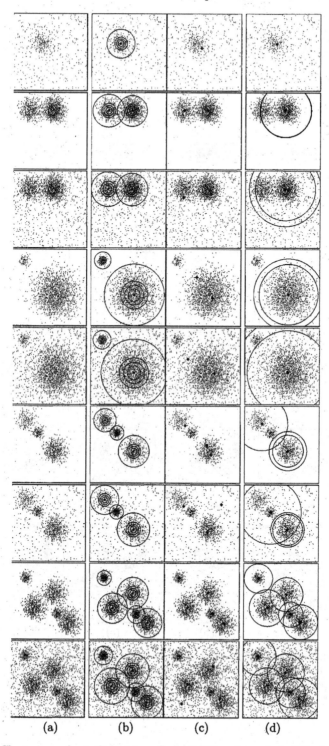

Fig. 8.9. Clustering spherical clusters: effect of different number of clusters, densities, sizes, and noise contamination rates: (a) original data set (b) results of UNC, (c) results of K-means with prespecified correct c, (d) results of PCM with prespecified correct c

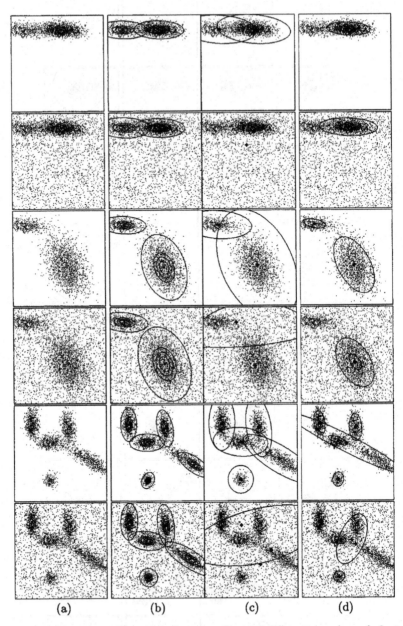

(a)   (b)   (c)   (d)

Fig. 8.10. Clustering ellipsoidal clusters: effect of different number of clusters, densities, sizes, and noise contamination rates: (a) original data set (b) results of UNC,(c) results of K-means with prespecified correct c, (d) results of PCM with prespecified correct c

Fig. 8.11. Image segmentation with UNC-scalar dispersion: (a) original image, (b) segmented image (non-core pixels in white) (c) boundary image

Fig. 8.12. Image segmentation with UNC-dispersion matrix: (a) original image, (b) segmented image (non-core pixels in white) (c) boundary image

into many smaller hyper-spherical clusters when Euclidean distance is used. This problem is less present in Fig. 8.12.

8.7 Conclusion

The Unsupervised Niche Clustering (UNC) algorithm is an unsupervised robust clustering technique based on genetic niching that is applicable to a variety of data mining applications. Using an appropriate density fitness measure, clusters in feature space are transformed into niches in the fitness landscape. The niche peaks, which represent the final cluster centers, are identified based on Deterministic Crowding (DC).

We eliminate the problem of crossover interactions in DC by allowing mating only between members within the same niche. For this purpose, we estimate the robust niche radii by using an iterative hill-climbing procedure coupled with the genetic optimization of the cluster centers. An upper bound on these radii is used to disqualify invalid solutions.

Our *one-step* local Baldwin Effect learning approach, enabled by a cumulative *memory* of past generations, is more efficient than traditional hybridizations of GA and hill-climbing procedures because the latter iteratively update the parameters within each generation. However, these updates cannot be used as starting points for the next generation due to arbitrary changes in the population distribution between generations, particularly in the early generations (no memory). Our approach has the advantage of not requiring more than a single step of updating the scales, and hence adapting the environment (fitness) to accelerate evolution. This makes convergence even faster compared to conventional hybrid algorithms.

UNC outperformed K-Means and to the Possibilistic C-Means clustering (PCM) algorithms on several examples with clusters varying in size, density, orientation, and noise contamination rates.

Compared to other unsupervised clustering techniques, UNC is a good candidate for clustering in many real world problems. This is because most other techniques are either object (feature) based, and thus necessitate the derivation of the optimal prototypes by differentiation to guarantee convergence to a *local* optimum, which can be impossible for most subjective and non-metric dissimilarity measures; or work on relational data consisting of a matrix of all pairwise dissimilarity values as in the case of the Graph based and Linkage type hierarchical clustering techniques. For these techniques, both the storage and computational cost can be prohibitively high (complexity in the order of $\mathcal{O}\left(N^2 \log_2\left(N\right)\right)$).

To summarize, this new approach to genetic clustering has the following advantages over previous methods: **(i)** It is *robust* in the presence of outliers and noise; **(ii)** it can *automatically determine the number of clusters*; **(iii)** because of the single cluster representation scheme used, *the size of the search space does not increase with the number of clusters or the number of data*;

(iv) It is *generic* enough that it can handle any type of distance/dissimiliarity measure and any type of input data (numerical, categorical, etc); and (v) it takes advantage of both evolutionary optimization for better *global* search, and *local Baldwin type learning* to accelerate and improve the accuracy of the search.

Acknowledgment

This work is supported by a National Science Foundation CAREER Award IIS-0133948 to O. Nasraoui.

References

8.1 Arabie, P., Hubert, L. J. (1996): An overview of combinatorial data analysis. In Arabie, P., Hubert, L.J., Soete, G. D., editors, Clustering and Classification, 5–63. World Scientific Pub., New Jersey

8.2 Bezdek, J.C. , Boggavarapu, S., Hall, L.O., Bensaid, A. (1994): Genetic algorithm guided clustering. In First IEEE conference on evolutionary computation, 1, 34–39, Orlando, Florida

8.3 Bezdek, J. C. (1981): Pattern recognition with fuzzy objective function algorithms. Plenum Press, New York

8.4 Bradley, P., Fayyad, U., Reina, C. (1998): Scaling clustering algorithms to large databases. In Proceedings of the 4th international conf. on Knowledge Discovery and Data Mining (KDD98)

8.5 Bradley, P., Fayyad, U., Reina, C. (1998): Scaling em (expectation-maximization) clustering to large databases. Technical Report MSR-TR-98-35, Microsoft Research

8.6 Babu, G. B., Murty, M.N. (1993): A near-optimal initial seed value selection in the k-means algorithm using a genetic algorithm. Pattern Recognition Letters, 14, 763–769

8.7 Bhuyan, J. N., Raghavan, V. V., Venkatesh, K. E (1991): Genetic algorithms for clustering with an ordered representation. In Fourth International Conference on Genetic Algorithms, 408–415

8.8 99]Blobworld Carson, C., Thomas, M., Belongie, S., Hellerstein, J. M., Malik., J. (1999): Blobworld: a system for region-based image indexing and retrieval. In VISUAL, 509–516, Amsterdam, The Netherlands

8.9 Deb, K., Goldberg, D. E. (1989): An investigation of niche and species formation in genetic function optimization. In 3rd Intl. Conf. Genetic Algsorithms, 42–50, San Mateo, CA

8.10 Duda, R., Hart., P. (1973): Pattern classification and scene analysis. Wiley Interscience, NY.

8.11 Dunn, J. C. (1994): A fuzzy relative of the isodata process and its use in detecting compact, well separated clusters. J. Cybernetics, 3, 32–57

8.12 Estivill-Castro, V., Yang, J. (2000): Fast and robust general purpose clustering algorithms. In Pacific Rim International Conference on Artificial Intelligence, 208–218

8.13 Fogel, L. J., Owens, A. J., Walsh, M. J. (1966): Artificial intelligence through simulated evolution. Wiley Publishing, New York

8.14 Fogel, D. B., Simpson, P. K. (1993): Evolving fuzzy clusters. In International Conference on Neural networks, pages 1829–1834, San Francisco, CA

8.15 Fukunaga, K. (1990): Introduction to statistical pattern recognition. Academic Press

8.16 Gustafson, E. E., Kessel, W. C. (1979): Fuzzy clustering with a fuzzy covariance matrix. In IEEE CDC, 761–766, San Diego, California

8.17 Goldberg, D.E. (1989): Genetic algorithms in search, optimization, and machine learning. Addison-Wesley, New York

8.18 Goldberg, D.E., Richardson., J. J. (1987): Genetic algorithms with sharing for multimodal function optimization. In 2nd Intl. Conf. Genetic Algsorithms, 41–49, Cambridge, MA

8.19 Guha, S., Rastogi, R., Shim, K. (1998): Cure: An efficient clustering algorithm for large data databases. In Proceedings of the ACM SIGMOD conference on Management of Data, Seattle Washington

8.20 Hall, L. O., Ozyurt, I. O., Bezdek, J. C. (1999): Clustering with a genetically optimized approach. IEEE Trans. Evolutionary Computations, 3, 103–112

8.21 Holland, J. H. (1975): Adaptation in natural and artificial systems. MIT Press

8.22 Hinton, G., Whitley, D. (1987): How learning can guide evolution. Complex Systems, 1, 495–502

8.23 Jolion, J. M., Meer, P., Bataouche, S. (1991): Robust clustering with applications in computer vision. IEEE Transactions on Pattern Analysis and Machine Intelligence, 13, 791–802

8.24 De Jong, K. A. (1975): An analysis of the behavior of a class of genetic adaptive systems. Doct. Diss., U. of Michigan., 36(10-5140B), 29–60

8.25 Krishnapuram, R., Freg, C. P. (1992): Fitting an unknown number of lines and planes to image data through compatible cluster merging. Pattern Recognition, 25

8.26 Krishnapuram, R., Keller, J. M. (1993): A possibilistic approach to clustering. IEEE Trans. Fuzzy Syst., 1, 98–110

8.27 Krishnapuram, R., Keller, J. M. (1994): Fuzzy and possibilistic clustering methods for computer vision. In Mitra, S., Gupta, M., Kraske, W., editors, Neural and Fuzzy Systems, 135–159. SPIE Institute Series

8.28 Lee, C. Y., Antonsson, E. K. (2000): Dynamic partitional clustering unsing evolution strategies. In 3rd Asia Pacific Conf. on simulated evolution and learning, Nagoya, Japan

8.29 MacQueen, J. (1967): Some methods for classification and analysis of multivariate observations. In Fifth Berkeley Symp. on Math. Statist. and Prob., 281–297, Berkeley, California, University of California Press

8.30 Mahfoud, S.W. (1992): Crowding and preselection revisited. In 2nd Conf. Parallel problem Solving from Nature, PPSN '92, Brussels, Belgium

8.31 Ng R.T., Han, J. (1994): Efficient and effective clustering methods for spatial data mining. In Proceedings of the VLDB conference, Santiago Chile

8.32 Nasraoui, O., Krishnapuram, R. (1996): A robust estimator based on density and scale optimization, and its application to clustering. In IEEE International Conference on Fuzzy Systems, 1031–1035, New Orleans

8.33 Nasraoui, O., Krishnapuram, R. (1997): Clustering using a genetic fuzzy least median of squares algorithm. In North American Fuzzy Information Processing Society Conference, Syracuse NY

8.34 Nasraoui, O., Krishnapuram, R. (2000): A novel approach to unsupervised robust clustering using genetic niching. In Ninth IEEE International Conference on Fuzzy Systems, 170–175, San Antonio, TX

8.35 Raghavan, V.V., K. Birchand, K. (1979): A clustering strategy based on a formalism of the reproductive process in a natural system. In Second International Conference on Information Storage and Retrieval, 10–22

8.36 Rousseeuw, P. J., Leroy, A.M. (1987): Robust regression and outlier detection. John Wiley & Sons, New York

8.37 Ruspini, E. (1969): A new approach to clustering. Information Control, **15**, 22–32

8.38 Whitley, D., Gordon, S., Mathias, K. (1994): Lamarckian evolution, the baldwin effect and function optimization. In Davidor, Y., Schwefel, H., Manner, R., editors, Parallel Problem Solving From Nature-PPSN III, 6–15. Springer Verlag

8.39 Zhang, T., Ramakrishnan, R., Livny, L. (1986): Birch: An efficient data clustering method for large databases. In ACM SIGMOD International Conference on Management of Data, 103–114, New York, NY, ACM Press

9. Evolutionary Computation in Intelligent Network Management

Ajith Abraham

Natural Computation Lab
Department of Computer Science
Oklahoma State University,USA
ajith.abraham@ieee.org

Abstract: Data mining is an iterative and interactive process concerned with discovering patterns, associations and periodicity in real world data. This chapter presents two real world applications where evolutionary computation has been used to solve network management problems. First, we investigate the suitability of linear genetic programming (LGP) technique to model fast and efficient intrusion detection systems, while comparing its performance with artificial neural networks and classification and regression trees. Second, we use evolutionary algorithms for a Web usage-mining problem. Web usage mining attempts to discover useful knowledge from the secondary data obtained from the interactions of the users with the Web. Evolutionary algorithm is used to optimize the concurrent architecture of a fuzzy clustering algorithm (to discover data clusters) and a fuzzy inference system to analyze the trends. Empirical results clearly shows that evolutionary algorithm could play a major rule for the problems considered and hence an important data mining tool.

9.1 Intrusion Detection Systems

Security of computers and the networks that connect them is increasingly becoming of great significance. Computer security is defined as the protection of computing systems against threats to confidentiality, integrity, and availability. There are two types of intruders: the external intruders who are unauthorized users of the machines they attack, and internal intruders, who have permission to access the system with some restrictions. The traditional prevention techniques such as user authentication, data encryption, avoiding programming errors and firewalls are used as the first line of defense for computer security. If a password is weak and is compromised, user authentication cannot prevent unauthorized use, firewalls are vulnerable to errors in configuration and ambiguous or undefined security policies. They are generally unable to protect against malicious mobile code, insider attacks and unsecured modems. Programming errors cannot be avoided as the complexity of the system and application software is changing rapidly leaving behind some exploitable weaknesses. Intrusion detection is therefore required as an additional wall for protecting systems [9.10, 9.15]. Intrusion detection is useful not only in detecting successful intrusions, but also provides important information for timely countermeasures [9.18, 9.19]. An intrusion is defined as any set of actions that attempt to compromise the integrity, confidentiality or availability of a resource. An attacker can gain access because of an error in the configuration of a system. In some cases it is possible to fool a system into giving access by misrepresenting oneself. An example is sending a TCP packet that has a forged source address that makes the packet appear to come from a trusted host. Intrusions may be classified into several types [9.19] .

— Attempted break-ins, which are detected by typical behavior profiles or violations of security constraints.
— Masquerade attacks, which are detected by atypical behavior profiles or violations of security constraints.
— Penetration of the security control system, which are detected by monitoring for specific patterns of activity.
— Leakage, which is detected by atypical use of system resources.
— Denial of service, which is detected by atypical use of system resources.
— Malicious use, which is detected by atypical behavior profiles, violations of security constraints, or use of special privileges.

The process of monitoring the events occurring in a computer system or network and analyzing them for sign of intrusions is known as Intrusion detection. Intrusion detection is classified into two types: misuse intrusion detection and anomaly intrusion detection.

— Misuse intrusion detection uses well-defined patterns of the attack that exploit weaknesses in system and application software to identify the intrusions. These patterns are encoded in advance and used to match against the user behavior to detect intrusion.
— Anomaly intrusion detection uses the normal usage behavior patterns to identify the intrusion. The normal usage patterns are constructed from the statistical measures of the system features, for example, the CPU and I/O activities by a particular user or program. The behavior of the user is observed and any deviation from the constructed normal behavior is detected as intrusion.

We have two options to secure the system completely, either prevent the threats and vulnerabilities which come from flaws in the operating system as well as in the application programs or detect them and take some action to prevent them in future and also repair the damage. It is impossible in practice, and even if possible, extremely difficult and expensive, to write a completely secure system. Transition to such a system for use in the entire world would be an equally difficult task. Cryptographic methods can be compromised if the passwords and keys are stolen. No matter how secure a system is, it is vulnerable to insiders who abuse their privileges. There is an inverse relationship between the level of access control and efficiency. More access controls make a system less user-friendly and more likely of not being used.

An Intrusion Detection system is a program (or set of programs) that analyzes what happens or has happened during an execution and tries to find indications that the computer has been misused. An Intrusion detection system does not eliminate the use of preventive mechanism but it works as the last defensive mechanism in securing the system. Data mining approaches are a relatively new technique for intrusion detection.

9.1.1 Intrusion Detection - a Data Mining Approach

Data mining is a relatively new approach for intrusion detection. Data mining approaches for intrusion detection was first implemented in Mining Audit Data for Automated Models for Intrusion Detection [9.14] . The raw data is first converted into ASCII network packet information which in turn is converted into connection level information. These connection level records contain within connection features like service, duration etc. Data mining algorithms are applied to this data to create models to detect intrusions. Data mining algorithms used in this approach are RIPPER (rule based classification algorithm), meta-classifier, frequent episode algorithm and association rules. These algorithms are applied to audit data to compute models that accurately capture the actual behavior of intrusions as well as normal activities.

The RIPPER algorithm was used to learn the classification model in order to identify normal and abnormal behavior [9.8] . Frequent episode algorithm and association rules together are used to construct frequent patterns from audit data records. These frequent patterns represent the statistical summaries of network and system activity by measuring the correlations among system features and sequential co-occurrence of events. From the constructed frequent patterns the consistent patterns of normal activities and the unique intrusion patterns are identified and analyzed, and then used to construct additional features. These additional features are useful in learning the detection model more efficiently in order to detect intrusions. RIPPER classification algorithm is then used to learn the detection model. Meta classifier is used to learn the correlation of intrusion evidence from multiple detection models and produce combined detection model. The main advantage of this system is automation of data analysis through data mining, which enables it to learn rules inductively replacing manual encoding of intrusion patterns. However, some novel attacks may not be detected.

Audit Data Analysis and Mining combines association rules and classification algorithm to discover attacks in audit data [9.5] . Association rules are used to gather necessary knowledge about the nature of the audit data as the information about patterns within individual records can improve the classification efficiency. This system has two phases, training phase and detection phase. In the training phase database of frequent item sets is created for the attack-free items from using only attack-free data set. This serves as a profile against which frequent item sets found later will be compared. Next a sliding-window, on-line algorithm is used to find frequent item sets in the last D connections and compares them with those stored in the attack-free database, discarding those that are deemed normal. In this phase classifier is also trained to learn the model to detect the attack. In the detection phase a dynamic algorithm is used to produce item sets that are considered as suspicious and used by the classification algorithm already learned to classify the item set as attack, false alarm (normal event) or as unknown. Unknown attacks are the ones which are not able to detect either as false alarms or as known attacks. This method attempts to detect only anomaly attacks.

9.1.2 Linear Genetic Programming (LGP)

Linear genetic programming is a variant of the GP technique that acts on linear genomes. Its main characteristics in comparison to tree-based GP lies in that the evolvable units are not the expressions of a functional programming language (like LISP), but the programs of an imperative language (like c/c ++) [9.7] . An alternate approach is to evolve a computer program at the machine code level, using lower level representations for the individuals. This can tremendously hasten up the evolution process as, no matter how an individual is initially represented, finally it always has to be represented as a piece of machine code, as fitness evaluation requires physical execution of the individuals. The basic unit of evolution here is a native machine code instruction that runs on the floating-point processor unit (FPU). Since different instructions may have different sizes, here instructions are clubbed up together to form instruction blocks of 32 bits each. The instruction blocks hold one or more native machine code instructions, depending on the sizes of the instructions. A crossover point can occur only between instructions and is prohibited from occurring within an instruction. However the mutation operation does not have any such restriction. One of the most serious problems of standard genetic programming is the convergence of the population. It has been often observed that unless convergence is achieved within certain number of generations, the system will never converge. Parallel populations or demes may possess different parameter settings that can be explored simultaneously, or they may cooperate with the same set of parameters, while each working on different individuals. In this research, we used circular movement of evolved programs among the demes, i.e., program movement can take place only between adjacent demes in the circle. Steady state genetic programming approach was used to manage the memory more effectively.

9.1.3 Decision Trees (DT) as Intrusion Detection Model

Intrusion detection can be considered as classification problem where each connection or user is identified either as one of the attack types or normal based on some existing data. Decision trees work well with large data sets. This is important as large amounts of data flow across computer networks. The high performance of Decision trees makes them useful in real-time intrusion detection. Decision trees construct easily interpretable models, which is useful for a security officer to inspect and edit. These models can also be used in the rule-based models with minimum processing [9.13] . Generalization accuracy of decision trees is another useful property for intrusion detection model. There will always be some new attacks on the system, which are small variations of known attacks after the intrusion detection models are built. The ability to detect these new intrusions is possible due to the generalization accuracy of decision trees.

9.1.4 Support Vector Machines (SVM)

Support Vector Machines have been proposed as a novel technique for intrusion detection. SVM maps input (real-valued) feature vectors into a higher dimensional feature space through some nonlinear mapping. SVMs are powerful tools for providing solutions to classification, regression and density estimation problems. These are developed on the principle of structural risk minimization. Structural risk minimization seeks to find a hypothesis h for which one can find lowest probability of error. The structural risk minimization can be achieved by finding the hyper plane with maximum separable margin for the data [9.22] .

Computing the hyper plane to separate the data points i.e. training a SVM leads to quadratic optimization problem. SVM uses a feature called kernel to solve this problem. Kernel transforms linear algorithms into nonlinear ones via a map into feature spaces. There are many kernel functions; some of them are Polynomial, radial basis functions, two layer sigmoid neural nets etc. The user may provide one of these functions at the time of training classifier, which selects support vectors along the surface of this function. SVMs classify data by using these support vectors, which are members of the set of training inputs that outline a hyper plane in feature space. The main disadvantage is SVM can only handle binary-class classification whereas intrusion detection requires multi-class classification.

9.1.5 Intrusion Detection Data

In 1998, DARPA intrusion detection evaluation program created an environment to acquire raw TCP/IP dump data for a network by simulating a typical U.S. Air Force LAN [9.16]. The LAN was operated like a real environment, but being blasted with multiple attacks. For each TCP/IP connection, 41 various quantitative and qualitative features were extracted. Of this database a subset of 494021 data were used for our studies, of which 20% represent normal patterns [9.12]. Different categories of attacks are summarized in Fig. 9.1. Attack types fall into four main categories:

1. **Probing: surveillance and other probing**

   Probing is a class of attacks where an attacker scans a network to gather information or find known vulnerabilities. An attacker with a map of machines and services that are available on a network can use the information to look for exploits. There are different types of probes: some of them abuse the computer's legitimate features; some of them use social engineering techniques. This class of attacks is the most commonly heard and requires very little technical expertise.

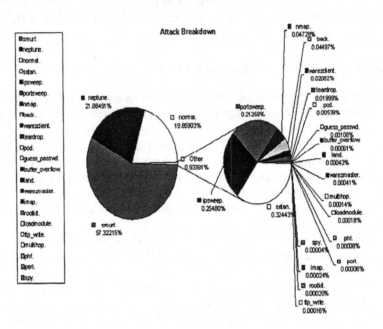

Fig. 9.1. Intrusion detection data distribution

2. **DoS: denial of service**

   Denial of Service (DoS) is a class of attacks where an attacker makes some computing or memory resource too busy or too full to handle legitimate requests, thus denying legitimate users access to a machine. There are different ways to launch DoS attacks: by abusing the computers legitimate features; by targeting the implementations bugs; or by exploiting the system's misconfigurations. DoS attacks are classified based on the services that an attacker renders unavailable to legitimate users.

3. **U2Su: unauthorized access to local super user (root) privileges**

   User to root (U2Su) exploits are a class of attacks where an attacker starts out with access to a normal user account on the system and is able to exploit vulnerability to gain root access to the system. Most common exploits in this class of attacks are regular buffer overflows, which are caused by regular programming mistakes and environment assumptions.

4. **R2L: unauthorized access from a remote machine**
   A remote to user (R2L) attack is a class of attacks where an attacker sends packets to a machine over a network, then exploits machine's vulnerability to illegally gain local access as a user. There are different types of R2U attacks; the most common attack in this class is done using social engineering.

Experimentation setup and results. We performed a 5-class classification. The (training and testing) data set contains 11982 randomly generated points from the data set representing the five classes, with the number of data from each class proportional to its size, except that the smallest class is completely included. The set of 5092 training data and 6890 testing data are divided in to five classes: normal, probe, denial of service attacks, user to super user and remote to local attacks. Where the attack is a collection of 22 different types of instances that belong to the four classes described earlier and the other is the normal data. The normal data belongs to class 1, probe belongs to class 2, denial of service belongs to class 3, user to super user belongs to class 4, remote to local belongs to class 5. Two randomly generated separate data sets of sizes 5092 and 6890 are used for training and testing the LGP, DT and SVM respectively.

Experiments using linear genetic programming. The settings of various linear genetic programming system parameters are of utmost importance for successful performance of the system. The population space has been subdivided into multiple subpopulation or demes. Migration of individuals among the subpopulations causes evolution of the entire population. It helps to maintain diversity in the population, as migration is restricted among the demes. Moreover, the tendency towards a bad local minimum in one deme can be countered by other demes with better search directions. The various LGP search parameters are the mutation and the crossover frequencies. The crossover operator acts by exchanging sequences of instructions between two tournament winners. After a trial and error approach, the following parameter settings were used to develop IDS.

Figs. 9.2, 9.3, 9.4, 9.5 and 9.6 demonstrates the growth in program length during 120,000 tournaments and the average fitness values for all the five classes. Test data classification accuracy is depicted in Table 9.2.

Experiments using support vector machines. Our trial experiments revealed that the polynomial kernel option often performs well on most of the data sets. Classification accuracies for the different types of attacks (test data) are depicted in Table 9.2

Experiments using decision trees. First a classifier is constructed using the training data and then testing data is tested with the constructed classifier to classify the data into normal or attack. Table 9.2 summarizes the results of the test data.

Table 9.1. Parameter settings for linear genetic programming

Parameter	Normal	Probe	DoS	U2Su	R2L
Population size	2048	2048	2048	2048	2048
Maximum no of tournaments	120000	120000	120000	120000	120000
Tournament size	8	8	8	8	8
Mutation frequency (%)	85	82	75	86	85
Crossover frequency (%)	75	70	65	75	70
Number of demes	10	10	10	10	10
Maximum program size	256	256	256	256	256

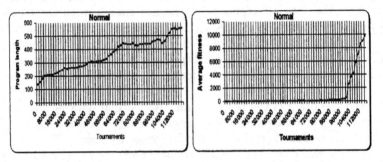

Fig. 9.2. Detection of normal patterns (a) growth in program length (b) average fitness

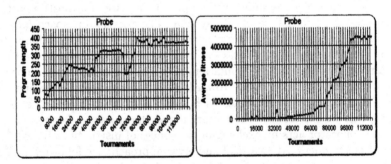

Fig. 9.3. Detection of probe (a) Growth in program length (b) average training fitness

Table 9.2. Parameter settings for linear genetic programming

Class type	Classification accuracy (%)		
	DT	SVM	LGP
Normal	99.64	99.64	99.73
Probe	99.86	98.57	99.89
DOS	96.83	99.92	99.95
U2R	68.00	40.00	64.00
R2L	84.19	33.92	99.47

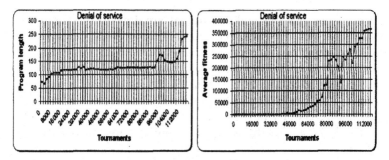

Fig. 9.4. Detection of DoS (a) growth in program length (b) average training fitness

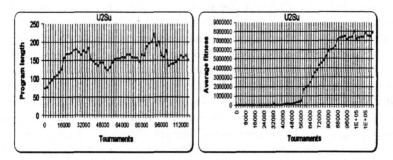

Fig. 9.5. Detection of U2Su (a) growth in program length (b) average training fitness

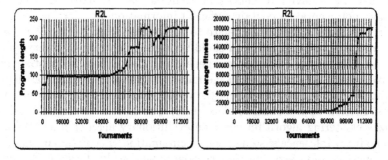

Fig. 9.6. Detection of R2L (a) growth in program length (b) average training fitness

9.1.6 Discussions

A number of observations and conclusions are drawn from the results illustrated in Table 9.1. LGP outperformed decision trees and support vector machines in terms of detection accuracies (except for one class). Decision trees could be considered as the second best, especially for the detection of U2R attacks. In some classes the accuracy figures tend to be very small and may not be statistically significant, especially in view of the fact that the 5 classes of patterns differ in their sizes tremendously. More definitive conclusions can only be made after analyzing more comprehensive sets of network traffic data.

9.2 Web usage Mining using Intelligent Miner (i-Miner)

The WWW continues to grow at an amazing rate as an information gateway and as a medium for conducting business. From the business and applications point of view, knowledge obtained from the Web usage patterns could be directly applied to efficiently manage activities related to e-business, e-services, e-education and so on. Web usage could be used to discover the actual contents of the web pages (text, images etc.), organization of the hyperlink architecture (HTML/XML links etc.) of different pages and the data that describes the access patterns (Web server logs etc.) [9.1, 9.20] . A typical Web log format is depicted in Fig. 9.7. When ever a visitor access the server it leaves the IP, authenticated user ID, time/date, request mode, status, bytes, referrer, agent and so on. The available data fields are specified by the HTTP protocol. In the case of Web mining, data could be collected at the server level, client level, proxy level or some consolidated data. These data could differ in terms of content and the way it is collected etc. The usage data collected at different sources represent the navigation patterns of different segments of the overall Web traffic, ranging from single user, single site browsing behavior to multi-user, multi-site access patterns. Web server log does not accurately contain sufficient information for inferring the behavior at the client side as they relate to the pages served by the Web server. Pre-processed and cleaned data could be used for pattern discovery, pattern analysis, Web usage statistics and generating association/ sequential rules.

We present a hybrid Web usage mining framework (i-miner) as depicted in Fig. 9.8 [9.4, 9.2]. by clustering the visitors and analyzing the trends using some function approximation algorithms. The hybrid framework optimizes a fuzzy clustering algorithm using an evolutionary algorithm and a Takagi-Sugeno fuzzy inference system using a combination of evolutionary algorithm and neural network learning. The raw data from the log files are cleaned and pre-processed and a fuzzy C means algorithm is used to identify the number of clusters. The developed clusters of data are fed to a Takagi-Sugeno fuzzy inference system to analyze the trend patterns. The if-then rule structures are

```
64.68.82.66 - - [17/May/2003:03:41:23 -0500] "GET /marcin HTTP/1.0" 404 318
192.114.47.54 - - [17/May/2003:03:41:33 -0500] "GET /~aa/isda2002/isda2002.html HTTP/1.1" 404 350
216.239.37.5 - - [17/May/2003:03:41:43 -0500] "GET /~ijcr/Vols/vol10no1.html HTTP/1.0" 200 4568
218.244.111.106 - - [17/May/2003:03:41:51 -0500] "GET /~aa/his/ HTTP/1.1" 404 332
64.68.82.18 - - [17/May/2003:03:42:15 -0500] "GET /~pdcp/cfp/cfpBookReviews.html HTTP/1.0" 304 -
212.98.136.62 - - [17/May/2003:03:43:21 -0500] "GET /cs3373/programs/pgm03.dat HTTP/1.1" 200 498
212.98.136.62 - - [17/May/2003:03:43:26 -0500] "GET /cs3373/programs/pgm04.html HTTP/1.1" 200 55722
212.98.136.62 - - [17/May/2003:03:43:38 -0500] "GET /cs3373/images/WaTor.gif HTTP/1.1" 200 39021
212.29.232.2 - - [17/May/2003:03:43:40 -0500] "GET /welcome.html HTTP/1.0" 200 5253
```

Fig. 9.7. Sample entries from a Web server access log

learned using an iterative learning procedure by an evolutionary algorithm and the rule parameters are fine-tuned using a backpropagation algorithm.

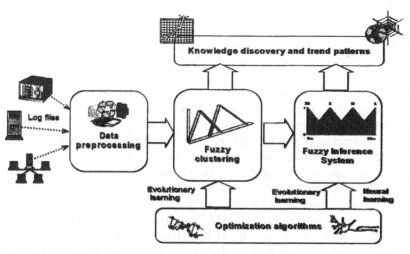

Fig. 9.8. i-Miner framework

The hierarchical distribution of the i-Miner is depicted in Fig. 9.9. The arrow direction depicts the speed of the evolutionary search. The optimization of clustering algorithm progresses at a faster time scale in an environment decided by the inference method and the problem environment.

9.2.1 Optimization of Fuzzy Clustering Algorithm

One of the widely used clustering methods is the fuzzy c-means (FCM) algorithm developed by Bezdek [9.6] . FCM partitions a collection of n vectors $x_i = 1, 2, ..., n$ into c fuzzy groups and finds a cluster center in each group such that a cost function of dissimilarity measure is minimized. To accommodate the introduction of fuzzy partitioning, the membership matrix U is allowed to have elements with values between 0 and 1.The FCM objective function takes the form

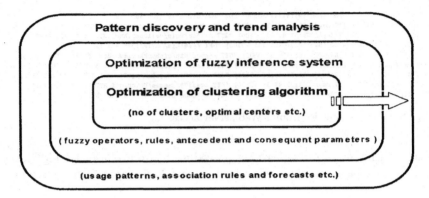

Fig. 9.9. Hierarchical architecture of i-Miner

$$J(U, c_1, ...c_c) = \sum_{i=1}^{c} Ji = \sum_{i=1}^{c} \sum_{j=1}^{n} u_{ij}^m d_{ij}^2, \tag{9.1}$$

where u_{ij}, is a numerical value between [0,1]; c_i is the cluster center of fuzzy group i; $d_{ij} = \|c_i - x_j\|$ is the Euclidian distance between i^{th} cluster center and j^{th} data point; and m is called the exponential weight which influences the degree of fuzziness of the membership (partition) matrix. Usually a number of cluster centers are randomly initialized and the FCM algorithm provides an iterative approach to approximate the minimum of the objective function starting from a given position and leads to any of its local minima [9.6] . No guarantee ensures that FCM converges to an optimum solution (can be trapped by local extrema in the process of optimizing the clustering criterion). The performance is very sensitive to initialization of the cluster centers. An evolutionary algorithm is used to decide the optimal number of clusters and their cluster centers. The algorithm is initialized by constraining the initial values to be within the space defined by the vectors to be clustered. A very similar approach is given in [9.11] .

9.2.2 Optimization of the Fuzzy Inference System

We used the EvoNF framework [9.3], which is an integrated computational framework to optimize fuzzy inference system using neural network learning and evolutionary computation. Solving multi-objective scientific and engineering problems is, generally, a very difficult goal. In these particular optimization problems, the objectives often conflict across a high-dimension problem space and may also require extensive computational resources. The hierarchical evolutionary search framework could adapt the membership functions (shape and quantity), rule base (architecture), fuzzy inference mechanism (T-norm and T-conorm operators) and the learning parameters of neural network learning algorithm. In addition to the evolutionary learning (global search) neural network learning could be considered as a local search technique to optimize the parameters of the rule antecedent/consequent parameters and the parameterized fuzzy operators. The hierarchical search could be formulated as follows: For every fuzzy inference system, there exist a global search of neural network learning algorithm parameters, parameters of the fuzzy operators, if-then rules and membership functions in an environment decided by the problem. The evolution of the fuzzy inference system will evolve at the slowest time scale while the evolution of the quantity and type of membership functions will evolve at the fastest rate. The function of the other layers could be derived similarly. Hierarchy of the different adaptation layers (procedures) will rely on the prior knowledge (this will also help to reduce the search space). For example, if we know certain fuzzy operators will work well for a problem then it is better to implement the search of fuzzy operators at a higher level. For fine-tuning the fuzzy inference system all the node functions are to be parameterized. For example, the Schweizer and Sklar's T-norm operator can be expressed as:

$$T(a, b, p) = [max\{0, (a^{-p} + b^{-p} - 1\})]^{-\frac{1}{p}}.$$

It is observed that

$$lim_{p \to 0} T(a, b, p) = ab$$

(9.2)

$$lim_{p \to \infty} T(a, b, p) = min\{a, b\},$$

which correspond to two of the most frequently used T-norms in combining the membership values on the premise part of a fuzzy if-then rule.

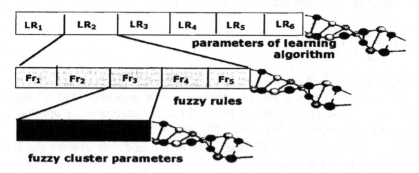

Fig. 9.10. Chromosome structure of the i-Miner

Chromosome modelling and representation. Hierarchical evolutionary search process has to be represented in a chromosome for successful modelling of the i-Miner framework. A typical chromosome of the i-Miner would appear as shown in Fig. 9.10 and the detailed modelling process is as follows.

Layer 1: The optimal number of clusters and initial cluster centers is represented this layer.

Layer 2: This layer is responsible for the optimization of the rule base. This includes deciding the total number of rules, representation of the antecedent and consequent parts. The number of rules grows rapidly with an increasing number of variables and fuzzy sets. We used the grid-partitioning algorithm to generate the initial set of rules [9.3]. An iterative learning method is then adopted to optimize the rules [9.9] . The existing rules are mutated and new rules are introduced. The fitness of a rule is given by its contribution (strength) to the actual output. To represent a single rule a position dependent code with as many elements as the number of variables of the system is used. Each element is a binary string with a bit per fuzzy set in the fuzzy partition of the variable, meaning the absence or presence of the corresponding linguistic label in the rule. For a three input and one output variable, with fuzzy partitions composed of 3,2,2 fuzzy sets for input variables and 3 fuzzy sets for output variable, the fuzzy rule will have a representation as shown in Fig. 9.5.

Layer 3: This layer is responsible for the selection of optimal learning parameters. Performance of the gradient descent algorithm directly depends on the learning rate according to the error surface. The optimal learning parameters decided by this layer will be used to tune the parameterized rule antecedents/consequents and the fuzzy operators. The rule antecedent/consequent parameters and the fuzzy operators are fine tuned using a gradient descent algorithm to minimize the output error

$$E = \sum_{k=1}^{N}(d_k - x_k)^2$$

where d_k is the kth component of the rth desired output vector and x_k is the kth component of the actual output vector by presenting the rth input vector to the network. All the gradients of the parameters to be optimized, namely the consequent parameters $\frac{\partial E}{\partial P_n}$ for all rules Rn and the premise parameters $\frac{\partial E}{\partial \sigma_i}$ and $\frac{\partial E}{\partial c_i}$ for all fuzzy sets F_i (σ and c represents the MF width and center of a Gaussian MF).

Once the three layers are represented in a chromosome C, and then the learning procedure could be initiated as follows:

1. Generate an initial population of N numbers of C chromosomes. Evaluate the fitness of each chromosome depending on the output error.

2. Depending on the fitness and using suitable selection methods reproduce a number of children for each individual in the current generation.
3. Apply genetic operators to each child individual generated above and obtain the next generation.
4. Check whether the current model has achieved the required error rate or the specified number of generations has been reached. Go to Step b.
5. End

Experimentation setup, training and performance evaluation. To demonstrate the efficiency of the proposed frameworks, Web access log data at the Monash University's Web site [9.17] were used for experimentations. We used the statistical/ text data generated by the log file analyzer from 01 January 2002 to 07 July 2002. Selecting useful data is an important task in the data pre-processing block. After some preliminary analysis, we selected the statistical data comprising of domain byte requests, hourly page requests and daily page requests as focus of the cluster models for finding Web users' usage patterns. It is also important to remove irrelevant and noisy data in order to build a precise model. We also included an additional input 'index number' to distinguish the time sequence of the data. The most recently accessed data were indexed higher while the least recently accessed data were placed at the bottom. Besides the inputs 'volume of requests' and 'volume of pages (bytes)' and 'index number', we also used the 'cluster information' provided by the clustering algorithm as an additional input variable. The data was re-indexed based on the cluster information. Our task is to predict (few time steps ahead) the Web traffic volume on a hourly and daily basis. We used the data from 17 February 2002 to 30 June 2002 for training and the data from 01 July 2002 to 06 July 2002 for testing and validation purposes.

Table 9.3. Parameter settings of i-Miner

Population size	30
Maximum no of generations	35
Fuzzy inference system	Takagi Sugeno
Rule antecedent membership functions	3 membership functions per input variable (parameterized Gaussian) linear parameters
Rule consequent parameters	
Gradient descent learning	10 epochs
Ranked based selection	0.50
Elitism	5 %
Starting mutation rate	0.50

The initial populations were randomly created based on the parameters shown in Table 9.1. We used a special mutation operator, which decreases the mutation rate as the algorithm greedily proceeds in the search space [9.9] . If the allelic value x_i of the i-th gene ranges over the domain a_i and b_i the mutated gene is drawn randomly uniformly from the interval $[a_i, b_i]$.

$$x_i = x_i + \triangle(t, b_i - x_i), if \omega = 0$$

(9.3)

$$= x_i + \triangle(t, b_i - a_i), if \omega = 1$$

where ω represents an unbiased coin flip $p(\omega = 0) = p(\omega = 1) = 0.5$, and

$$\triangle(t, x) = x \left(1 - \gamma^{\left(1 - \frac{t}{t_{max}}\right)^b}\right)$$

defines the mutation step, where γ is the random number from the interval [0,1] and t is the current generation and t_{max} is the maximum number of generations. The function computes a value in the range [0,x] such that the probability of returning a number close to zero increases as the algorithm proceeds with the search. The parameter b determines the impact of time on the probability distribution $\triangle$ over [0,x]. Large values of b decrease the likelihood of large mutations in a small number of generations. The parameters mentioned in Table 9.1 were decided after a few trial and error approaches. Experiments were repeated 3 times and the average performance measures are reported. Figs. 9.11 and 9.12 illustrates the meta-learning approach combining evolutionary learning and gradient descent technique during the 35 generations.

Table 9.4 summarizes the performance of the developed i-Miner for training and test data. Performance is compared with the previous results [23] wherein the trends were analyzed using a Takagi-Sugeno Fuzzy Inference System (ANFIS) learned using neural network learning techniques and Linear Genetic Programming (LGP). The Correlation Coefficient (CC) for the test data set is also given in Table 9.4.

Figs. 9.13 and 9.14 illustrate the actual and predicted trends for the test data set. FCM approach created 9 data clusters for daily traffic according to the input features compared to 7 data clusters (Fig. 9.15) for the hourly requests.

Table 9.4. Performance of the different paradigms

Method	Period					
	Daily (1 day ahead)			Hourly (1 hour ahead)		
	RMSE		CC	RMSE		CC
	Train	Test		Train	Test	
i-Miner	0.0044	0.0053	0.9967	0.0012	0.0041	0.9981
TKFIS	0.0176	0.0402	0.9953	0.0433	0.0433	0.9841
LGP	0.0543	0.0749	0.9315	0.0654	0.0516	0.9446

The 35 generations of meta-learning approach created 62 if-then Takagi-Sugeno type fuzzy rules (daily traffic trends) and 64 rules (hourly traffic

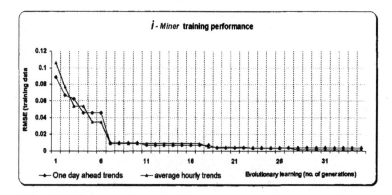

Fig. 9.11. Meta-learning performance (training) of i-Miner

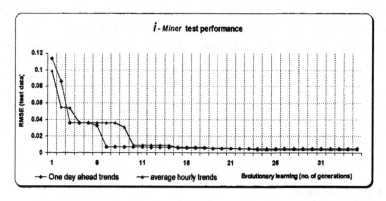

Fig. 9.12. Meta-learning performance (testing) of i-Miner

trends). Fig. 9.16 depicts the hourly visitor information according to domain names from an FCM cluster. Fig. 9.17 illustrates the volume of visitors in each FCM cluster according to the day of access.

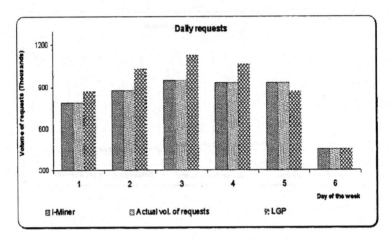

Fig. 9.13. Test results of the daily trends for 6 days

9.2.3 Discussions

Recently Web usage mining has been gaining a lot of attention because of its potential commercial benefits. Empirical results show that the proposed i-Miner framework seems to work very well for the problem considered. i-Miner framework gave the overall best results with the lowest RMSE on test error and the highest correlation coefficient. An important disadvantage of i-Miner is the computational complexity of the algorithm. When optimal performance is required (in terms of accuracy and smaller structure) such algorithms might prove to be useful as evident from the empirical results. In i-Miner evolutionary algorithm was used to optimize the various clustering and fuzzy inference system parameters. It is interesting to note that even LGP as a function approximator could pick up the trends accurately.

9.3 Conclusions

In this chapter, we have illustrated the importance of evolutionary algorithms for the two network management related problems. For real time intrusion detection systems LGP would be the ideal candidate as it could be manipulated at machine code level. Experiments using the Web data has revealed

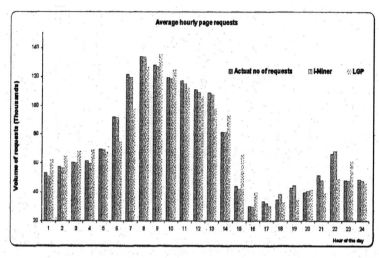

Fig. 9.14. Test results of the average hourly trends for 6 days

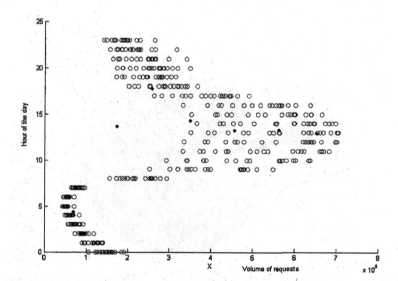

Fig. 9.15. Evolutionary FCM clustering: hour of the day and volume of requests. The dark dots indicate the cluster centers

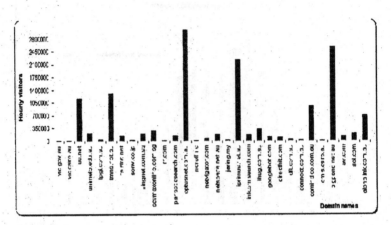

Fig. 9.16. Hourly visitor information according to the domain names from an FCM cluster

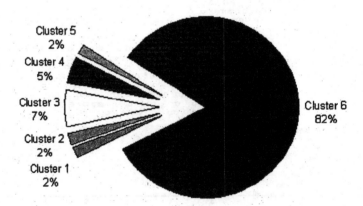

Fig. 9.17. Clustering of visitors based on the day of access from an FCM cluster

the importance of the optimization of fuzzy clustering algorithm and fuzzy inference system. Among the various trend analysis algorithms considered, LGP has again shown the capability as a robust function approximator.

Acknowledgements

Author wishes to thank Ms. Sandhya Peddabachigari (Oklahoma State University, USA), Mr. Vivek Gupta (IIT Bombay), Mr. Srinivas Mukkamala (New Mexico Tech, USA) and Ms. Xiaozhe Wang (Monash University, Australia) for all the valuable contributions during the different stages of this research.

References

9.1 Abraham, A., Ramos, V. (2003): Web usage mining using artificial ant colony clustering and genetic programming, 2003 IEEE Congress on Evolutionary Computation (CEC2003), Australia, IEEE Press, 1384-1391

9.2 Abraham, A. (2003): Business intelligence from web usage mining, Journal of Information & Knowledge Management (JIKM), World Scientific Publishing Co., Singapore, **2**

9.3 Abraham, A. (2002): EvoNF: A framework for optimization of fuzzy inference systems using neural network learning and evolutionary computation, In Proceedings of 17th IEEE International Symposium on Intelligent Control, IEEE Press, 327-332

9.4 Abraham, A. (2003): i-Miner: a web usage mining framework using hierarchical intelligent systems, The IEEE International Conference on Fuzzy Systems FUZZ-IEEE'03, IEEE Press, 1129-1134

9.5 Barbara, D., Couto, J., Jajodia, S., Wu, N. (2001): ADAM: a testbed for exploring the use of data mining in intrusion detection. SIGMOD Record, **30**, 15-24

9.6 Bezdek, J. C. (1981): Pattern recognition with fuzzy objective function algorithms, New York: Plenum Press.

9.7 Brameier, M., Banzhaf, W. (2001): A comparison of linear genetic programming and neural networks in medical data mining. Evolutionary Computation," IEEE Transactions on, **5**, 17-26

9.8 Cohen, W. (199): Learning trees and rules with set-valued features, American Association for Artificial Intelligence (AAAI).

9.9 Cordón, O., Herrera, F., Hoffmann, F., Magdalena, L. (2001): Genetic fuzzy systems: evolutionary tuning and learning of fuzzy knowledge bases, World Scientific Publishing Company, Singapore.

9.10 Denning, D. (1987): An intrusion-detection model, IEEE Transactions on Software Engineering, **13**, 222-232

9.11 Hall, L.O., Ozyurt, I.B., Bezdek J. C. (1999): Clustering with a genetically optimized approach, IEEE Transactions on Evolutionary Computation, **3**, 103-112

9.12 KDD cup (1999): intrusion detection data set: <http://kdd.ics.uci.edu/databases/kddcup99/kddcup.data_10_percent.gz>

9.13 Brieman, L., Friedman, J., Olshen, R., Stone, C. (198): Classification of regression trees. Wadsworth Inc.

9.14 Lee, W., Stolfo, S., Mok, K. (1999): A data mining framework for building intrusion detection models. In Proceedings of the IEEE Symposium on Security and Privacy.

9.15 Luo, J., Bridges, S. M. (2000): Mining fuzzy association rules and fuzzy frequency episodes for intrusion detection. International Journal of Intelligent Systems, John Wiley & Sons, 15, 687-704

9.16 MIT Lincoln Laboratory.
<http://www.ll.mit.edu/IST/ideval/>

9.17 Monash University Web site:
<http://www.monash.edu.au>

9.18 Mukkamala, S., Sung, A. H., Abraham A. (2003): Intrusion detection using ensemble of soft computing paradigms. Third International Conference on Intelligent Systems Design and Applications, Intelligent Systems Design and Applications, Advances in Soft Computing, Springer Verlag, Germany, 239-248

9.19 Peddabachigari, S., Abraham, A., Thomas, J. (2003): Intrusion detection systems using decision trees and support vector machines, International Journal of Applied Science and Computations, USA

9.20 Srivastava, J., Cooley, R., Deshpande, M., Tan, P. N. (2000): Web usage mining: discovery and applications of usage patterns from web data, SIGKDD Explorations, 1, 12-23

9.21 Summers, R. C. (1997): Secure computing: threats and safeguards. McGraw Hill, New York

9.22 Vapnik, V. N. (2002): The nature of statistical learning theory. Springer in Wang, X., Abraham, A., Smith, K. A. Soft computing paradigms for web access pattern analysis. Proceedings of the 1st International Conference on Fuzzy Systems and Knowledge Discovery, 631-635

10. Genetic Programming in Data Mining for Drug Discovery

W. B. Langdon and S. J. Barrett

Data Exploration Sciences
GlaxoSmithKline,
Research and Development,
Greenford, Middlesex,
UK.
http://www.cs.ucl.ac.uk/staff/W.Langdon

Abstract: Genetic programming (GP) is used to extract from rat oral bioavailability (OB) measurements simple, interpretable and predictive QSAR models which both generalize to rats and to marketed drugs in humans. Receiver Operating Characteristics (ROC) curves for the binary classifier produced by machine learning show no statistical difference between rats (albeit without known clearance differences) and man. Thus evolutionary computing offers the prospect of *in silico* ADME screening, e.g. for "virtual" chemicals, for pharmaceutical drug discovery.

The discovery, development and approval of a new drug treatment is a major undertaking (see Table 10.1). Only a small fraction of the drug discovery projects undertaken eventually lead to a successful medicine. Even successful programmes can take in the region of 12–15 years.

The discovery of new chemical entities with appropriate biological activity is a multi-stage and iteratively focussed search process in which many thousands of chemicals are measured firstly for primary activity against some (often novel) disease/therapy related target. The initial active subset subsequently becomes slimed down to select suitable candidates for use within the human body and worthy of expensive further development. The drug discovery process can be thought of as a funnel. The mouth of the funnel is wide and covers many diverse molecules. Gradually the funnel narrows and the later stages concentrate upon fewer more similar molecules.

As the directed discovery cycle continues, the criteria for progression become more stringent and complex. This means smaller numbers or classes of molecules are passed to the succeeding stages. Initially molecules need only show some hint of activity against the target in relatively cheap *in vitro* tests. Later (early development) stages progress to more expensive and time consuming *in vitro* and *in vivo* measurements. *In vivo* measurements are required to demonstrate good bodily Absorption, Distribution, Metabolism, Excretion and Toxicity (ADMET) properties.

ADMET testing includes satisfying aspects relating to: 1) metabolism by, or inhibition of, critical metabolic enzymes (such as cytochrome P450) and 2) the molecule's ability to reach and stay in areas of the body required to enable sufficient drug effect to occur before its metabolism/excretion. (These

Table 10.1. Discovery and early development cascade. At each successive stage in the cascade there is some increase in knowledge, but although the data becomes of higher quality it relates to a smaller (more specialized) chemical space as the decision about which chemicals to take forward becomes focussed to a smaller number of molecular classes and more complex as more factors are introduced

Exploratory Screening
High/ultra-High Throughput Screening HTS/u-HTS

Primary screening wells contain a single concentration of test chemical and the target, together with reagents for a "bio-assay". The assay is designed to show if the chemical directly binds to the target at all, or can promote some bioactivity via interaction with the target. Many tens of thousands or even hundreds of thousands of very diverse chemicals are tested.

IC50/EC50 and early selectivity assays. More refined measurements of primary target binding/potency involve testing at a number of chemical concentrations to determine the concentration that is needed to reach 50% of maximum inhibition/activity.

Another set of assays (also known as initial selectivity assays) are designed to test non-target specific binding/activity for avoiding other (unwanted) effects. Thousands to many tens of thousands of chemicals are tested, depending upon earlier "hit" rates, the number of molecular classes with promise for required activity or modifiability, their collective content (the initial Structure-Activity Relations information, "early SAR") and the initial importance of specific selectivity.

"Early Lead"/"Back-up". Selection of promising molecular classes with the necessary potency and selectivity and which its feasible to mass manufacture. Early lead/back-up chemicals should allow for their "optimization" as a drug.

Lead Optimisation

Chemical Programme of Modification

Using "*in silico*" virtual compound screening (i.e. selecting promising chemicals based on computer models) and/or more traditional QSAR/"rational" library design methods and combinatorial chemistry techniques many thousands of chemicals similar to the lead molecule class are identified and made in the hope they will have analogous and maybe improved properties.

Molecular class-focussed SAR screening with 1st and 2nd assays

Pharmacological characterization to improve potency and/or selectivity.

Initial key and "scale-feasible" ADMET related testing *in vitro*

Permeability, p450 interactions, solubility, etc. Results are feed back to the "Chemical Modification" stage, leading to an iterative cycle of chemical design, synthesis and testing.

"Development Candidate" with good potency, selectivity and initial ADMET

Further *in vivo* and more realistic and extensive ADMET and pharmacokinetic testing, including bioavailability measurements.

Again results are fed back to an increasingly more fine-tuned "Chemical Modification" stage.

Finally a compound is fit to be forwarded to first time in man (toxicity, dose-ranging) studies and subsequent clinical trials, plus supporting knowledge to be used in developing formulations/treatments.

properties are collectively known as good pharmacokinetics and bioavailability.)

Even an approximate *in silico* (computational) method that can be applied at an earlier step is very useful. Since it can be used by medicinal chemists to assist them to decide which molecules to progress. The time and expense required for *in vivo* testing and its inherently low through-put nature make measuring ADMET properties of the many thousands of chemicals at the mouth of the funnel infeasible. However poor ADMET characteristics cause a high failure-rate of molecules in the later stages of drug discovery. An approximate screen helps ensure better quality molecules advance to ADMET measurement from earlier stages. That is, *in silico* screening helps to ensure the later stages of the drug discovery funnel are not clogged with chemicals which will ultimately have to be rejected due to their poor ADMET characteristics.

We illustrate the use of genetic programming in drug discovery by using it to evolve simple, biologically interpretable *in silico* models of bioavailability. Section 10.4 summarizes GP, while Section 10.5 introduces ROC curves. Sections 10.6–10.7 and 10.8 describes the rat and human drug data sets and method. Our results (10.9), particularly Fig. 10.15 on page 231, are discussed in Section 10.10 before we conclude with Section 10.11. However first we give the background of using computational, data mining and machine learning techniques in drug discovery (Sections 10.1 and 10.2) and describe bioavailability (Section 10.3).

10.1 Computational Drug Discovery

There is a long history of efforts to improve earlier decision-making in the drug discovery process. This has largely involved Quantitative Structure-Activity Relationship (QSAR) modelling employing traditional multivariate statistical techniques (see for example [10.1], while [10.2] gives a current review) and some quite fundamental indicators have arisen [10.3].

QSAR models link chemical's structure to their pharmacological activity. They are applicable to both library design (i.e. selection of which chemicals to keep in a drug discovery "library" or warehouse) and "virtual screening" (see [10.4]). The predictive performance of QSAR models is typically highly related to the number and diversity of chemicals that are used in modelling. This in turn is curtailed by the extent of biological testing that has been done. (Mostly the data has not been gathered directly to support the modelling process.)

In more recent years "machine learning" approaches have increasingly been experimented with (see [10.5, 10.6]) and applied in this area. Although evolutionary computing techniques, principally genetic algorithms, have been used for some while [10.7], e.g. in library design [10.8], newer paradigms such as genetic programming have only more recently been experimented with

[10.9, 10.10] including the prediction of specific properties [10.11, 10.12, 10.13, 10.14, 10.15]. Although there is interest in other computational techniques [10.16].

10.2 Evolutionary Computing for Drug Discovery

Evolutionary computing and in particular genetic programming [10.17, 10.18] is increasingly being used to analyse bioinformatics data. For example recently two workshop series have started (BioGEC in the USA [10.19, 10.20] and EvoBIO in Europe [10.21]). Recent work also includes Kell [10.22], Moore [10.23], Iba [10.24] and Koza [10.25].

In the last few years we have used genetic programming in a series of data mining and modelling experiments. We used GP to fuse together models generated by other machine learning techniques (artificial neural networks, decision trees and naive Bayes [10.26, 10.27, 10.28]). This technique has been used for predicting inhibition of human cytochrome P450 2D6 (an important enzyme involved with the metabolism of many drugs) and compounds which might be starting points in drug based disease treatments [10.12, 10.13]. In a recent comparison, GP was shown to evolve models that best extrapolated from the available training data [10.14] as well as being understandable. Initial experiments have also used GP for both feature selection and model building of gene expression data [10.29].

10.3 Oral Bioavailability

The preferred method of introducing a drug into the body is for the patient to swallow it by mouth (orally). Hence a very important QSAR problem is the prediction of human oral bioavailability, %F. (%F is the percentage of an orally-administered dose reaching the blood stream.) Although some progress has been made, this task has proven particularly difficult. Typically false positive rates are quite high [10.4, 10.30, 10.31, 10.32] due to :

- The complex nature of processes underlying bioavailability. Bioavailability represents, in essence, the overall product of retained drug integrity within the body (e.g. avoiding high rates of digestion, metabolism in the liver and excretion via the kidneys). And the drug's ability to pass through bodily barriers while retaining its activity until the site of action (the target) is reached. Bioavailability is also a dose-dependent feature for many drugs. Obviously there are many physical and chemical processes involved.
- Modelling bioavailability is also hard due to the restricted set of classes of molecules for which %F measurements are available. This is because oral bioavailability is usually only measured late in the development process. So %F in man is normally only available for successfully marketed drugs or for proto-drugs which failed near the end of the drug discovery process.

Earlier efforts at examining molecular properties in relation to human oral bioavailability were naturally restricted to human data, but more recently progress has been made in identifying the importance of certain molecular properties from more voluminous rat data [10.33, 10.34].

10.4 Genetic Programming

Essentially genetic programming evolves a population of initially randomly created programs using a fitness function to select the better ones to be parents for children in succeeding generations. The processes of sexual recombination and mutation are used to ensure, that while children have some similarity with previous generations, they are different from them. Using "survival of the fittest" better programs are produced.

It is common in GP to represent programs as parse trees (Lisp S-expressions). (Fig. 10.8 an page 226 contains an example.) Parse trees have the great advantage that if one starts with two syntactically correct programs they can be used as parents of new children by exchanging subtrees between them (crossing over) and automatically the new programs will also be syntactically correct. Similarly various pruning and grafting operations can be implemented to mutate trees in such a way that they also give syntactically correct offspring. With a little practise, one can become adept at interpreting small trees and so extracting biological inferences from automatically created models.

GP can be thought of the use of genetic algorithms [10.35] to search the space of programs for one of the huge number of programs which satisfy some user requirement. Here the computer programs are all functional models of chemical properties. In addition to making accurate predictions of chemical properties, we will also want our models to be readily interpretable. It is difficult to quantify how interpretable a model is to a particular chemist or biologist, however, as a first step, it is reasonable to prefer smaller models. For an introduction into the long history of research into this aspect of genetic programming see [10.36].

10.5 Receiver Operating Characteristics

This section gives some of the background of Receiver Operating Characteristics (ROC) curves, while Section 10.5.1 uses ROC curves to explain why fitness $= \frac{1}{2}$True Positive rate $+ \frac{1}{2}$True Negative rate may be a good choice. That is, to explain why we use the average error rate rather than the error rate directly. (Note the two need not be the same when there are different numbers of training examples in each class. This is quite common.)

ROC curves are a good way to show the trade off a classifier makes between catching positive examples and raising false alarms [10.37]. Fig. 10.9

(page 227) shows some ROC curves. ROC curves plot a classifier's true positive rate (i.e. fraction of positive examples correctly classified) against its false positive rate (i.e. fraction of negative examples which it gets wrong). All ROC curves lie in the unit square and must pass through two points 0,0 and 1,1. The origin 0,0 corresponds to when the classifiers sensitivity is so low that it always says no. I.e. it never detects any positive examples. While 1,1 means the classifier is so sensitive that it always says yes. I.e. it never makes a mistake on positive examples, but gets all negative cases wrong. A classifier that randomly guesses has an ROC which lies somewhere along the diagonal line connecting 0,0 and 1,1. A better than random classifier, has an ROC above the diagonal. Since an ideal classifier detects all positive cases but does not raise any false alarms its ROC curve goes through 1,0 (the top left corner of unit square).

A worse than random classifier has an ROC curve lying below the diagonal. It can be converted into better than random by inverting its output. This has the effect of rotating its ROC curve by 180 degrees, so that the ROC curve now lies above the diagonal.

Scott [10.38] suggests that the "Maximum Realizable Receiver Operating Characteristics" for a combination of classifiers is the convex hull of their individual ROCs, cf. also [10.39]. In [10.38] Scott proves a nice result: it is always possible to form a composite classifier whose ROC lies at any chosen point within the convex hull of the original classifier's ROC. (See Figs. 10.1 and 10.2.) Since Scott's classifier is formed by random combination of the outputs, it is not acceptable where decisions have to justified individually, rather than on average across a group. E.g. some medical applications. Moreover the convex hull is not always the best that can be achieved [10.40]. Indeed we have shown GP can in some cases do better, including on Scott's own benchmarks [10.27] and several real world pharmaceutical classification tasks [10.11].

It can be shown that the overall discriminative ability of a classifier (as measured by its Wilcox statistic) is equal to the area under its ROC curve [10.41]. Thus we have used the area under the ROC (AUROC) to decide which classifiers to allow to breed and have children in the next generation. That is, we have used AUROC as the fitness measure.

10.5.1 Simple use of ROC as the Objective to be Maximized

It is not necessary to measure each Receiver Operating Characteristics curve completely during evolution. Instead some computational savings can be made by basing the fitness function on measuring only one point on the ROC. We can still form the convex hull of the ROC, but now it is always a quadrilateral (see Fig. 10.3). Given a little geometry, it can be shown that the area of the quadrilateral (i.e. the AUROC of the convex hull) is $\frac{1}{2}$ True Positive rate $+ \frac{1}{2}$ True Negative rate. This gives a very simple formula

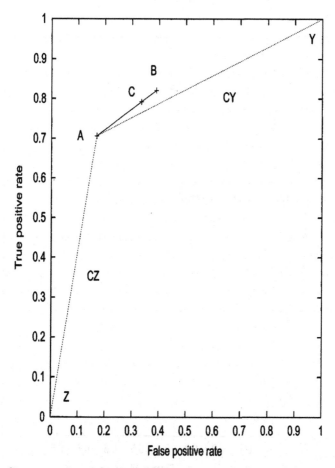

Fig. 10.1. Given two classifiers (A and B) a composite classifier (C) can always be formed by returning the result of A a fixed fraction of the time and the prediction given by B otherwise. The receiver operating characteristics of C will lie on a straight line connecting A and B. By combining with the classifier which always says no (Z) a composite CZ can be constructed between a real classifier and the origin (Z). Similarly, a classifier which always says yes (Y) can always be used to give a classifier (CY) between a real classifier and the 1,1 corner

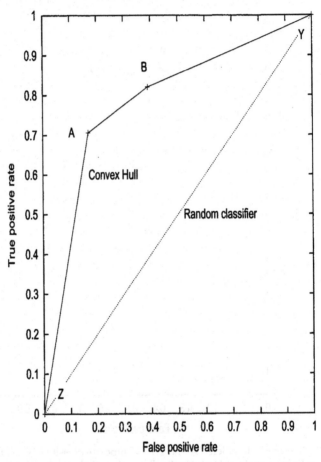

Fig. 10.2. A composite classifier can always be formed which will have an ROC lying between the convex hull of all available classifiers and the diagonal line between 0,0 and 1,1. (Genetic programming and other techniques can sometimes fuse classifiers to yield improved classifiers which lie above the convex hull)

for fitness calculation which automatically takes into account class imbalances.

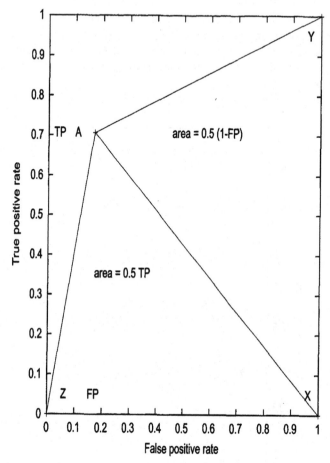

Fig. 10.3. With only one measured point on the ROC, the convex hull reduces to a quadrilateral (ZAYX). ZAYX is composed of triangles ZAX and XAY. ZAX's area is $\frac{1}{2}$TP. XAY's area is $\frac{1}{2}(1 - \text{FP})$

Note again geometry tells us that two points that are equally distant from the diagonal crossing the ROC square from 0,0 to 1,1 will give rise to quadrilaterals with the same area and hence have the same fitness. Another way of looking at this is to consider the cost associated with the classifier.

We assume the costs can be represented as α = cost of a false positive (false alarm) and β = cost of missing a positive (false negative). Let p be the proportion of positive cases. Then the average cost of classification at point x, y in the ROC space is $(1 - p)\alpha x + p\beta(1 - y)$.

Lines of equal cost are parallel and straight. Their gradient is $\alpha/\beta \times (1-p)/p$. If the cost of error on the two classes are equal ($\alpha = \beta$) and 50% are positive ($p = 0.5$), the gradient is 1 and the lines are at 45 degrees, i.e. parallel to the diagonal. Note that our GP fitness function ($\frac{1}{2}$True Positive rate + $\frac{1}{2}$True Negative rate) also rewards equally points that are equally distant from the diagonal. In other words it treats errors on both classes as if they were being equally important. It is a simple alternative to error rate, and it deals with the common case that more training data is available for one class than the other.

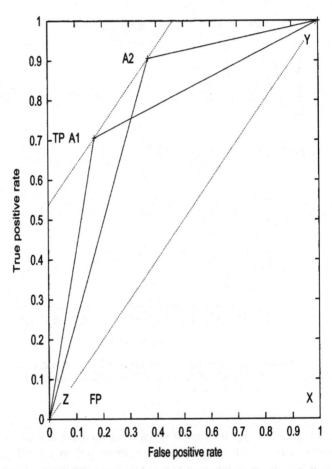

Fig. 10.4. With only one measured point on the ROC, the convex hull reduces to a quadrilateral (ZAYX) whose area is $0.5\text{TP} + 0.5(1 - \text{FP})$. Since A_1 and A_2 are the same distance from the diagonal ZY, the triangles ZA_1Y and ZA_2Y have the same area. Thus the quadrilaterals ZA_1YX and ZA_2YX have the same area

10.6 The Bioavailability Data

In high throughput screening (HTS) [10.11, 10.12, 10.13] and IC50 [10.14] experiments the chemical properties of many thousands of chemicals were measured in solution. In contrast, oral bioavailability is measured in living organisms. This severely limits the number of measurements. In fact only 481 data points on human subjects were initially available. These are data for marketed drugs.

The available data were randomly partitioned into 321 to be used for training and 160 held back as a holdout set. I.e. to assess how well the evolved models perform on data which they were not trained on.

The chemicals selected for measurement are naturally a highly biased sample. The sample consists only of chemicals which made it through the drug discovery process. There are two things we would like to know about any predictive classifier; how well it will work on chemicals like those on which it was trained and secondly (and much more difficult), how well will it extrapolate outside the training domain.

In addition to the training data we also had access to two further data sets. A further 124 human records, and data from rats. There was almost no overlap between the drugs in the human data sets and the chemicals in the rat data set. However the rat data set was naturally more extensive. Before using the rat data, chemicals known to have a specific bioavailability difference (related to "clearance") in humans to that in rats where excluded, leaving 2013 chemicals. Fig. 10.5 shows, as expected, there are systematic differences between the rat and the human data.

10.7 Chemical Features

An implicit modelling goal is that the model should be readily applicable to novel chemicals. Indeed we may wish to use the evolved model to predict the chemical properties of chemicals that have yet to be synthesised. I.e. "virtual chemicals" which exist only in the computer so far. Therefore the models must be based on chemical formulae, rather than three dimensional shapes (conformations). Since chemistry (particularly biochemistry) is inherently three dimensional this is a fundamental restriction. Nonetheless, as we shall see, predictions can still be made.

Chemical formulae can be viewed as graphs with labelled nodes (the atoms) and labelled edges (the bonds) (see Fig. 10.6). However it can be inconvenient to work with such graphs. Instead it is common practise in Chemoinformatics not to work from the chemical formulae directly but instead to precalculate chemical "features". The pharmaceutical industry has considerable experience with designing features. Simple features include, the presence or absence of charged atoms, aromatic rings, specific groups and metallic atoms.

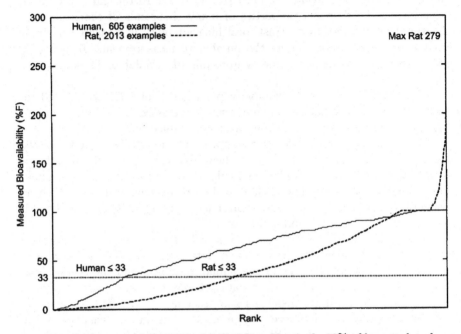

Fig. 10.5. Human and rat bioavailability data Note only 19% of human data has a bioavailability of 33 or below, while 47% of the rat data set is in class 0

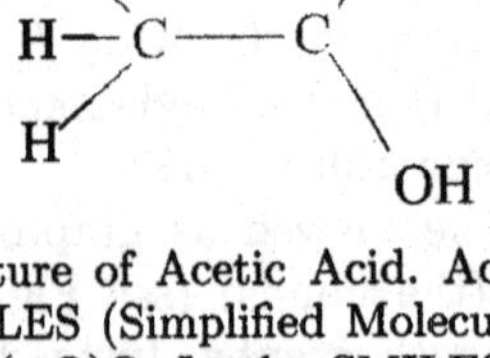

Fig. 10.6. Chemical structure of Acetic Acid. Acetic Acid's chemical formula is CH_3CO_2H. While its SMILES (Simplified Molecular Input Line Entry Specification) representation is CC(=O)O. In the SMILES representation all hydrogen atoms are omitted, branches are shown with parenthesis and double bonds with "$=$"

A total of 83, numerical and categorical, chemical features from a diverse array of families (electronic, structural, topological/shape, physico-chemical, etc.) were computed for each chemical, starting from a SMILES[1] representation of it's primary chemical structure (2-d chemical formula). Of the 83 features, all but 7 had previously been used in when modelling chemical interaction with a P450 cell wall enzyme [10.14].

10.8 Genetic Programming Configuration

The genetic programming system is deliberately simple. For example the GP uses a single type. So categorical and integer as well as continuous measurements and features are converted to single precision floating point numbers before the GP is run. The GP is summarized in Table 10.2.

10.8.1 Function set

The functions used for combining feature values and numerical values are the four binary arithmetic functions $(+-*/)$ and a four input "if". Since functions are initially used randomly, division is "protected". This means division by zero is trapped and the value 1 is returned rather than attempting to perform division by zero.

IFLTE evaluates its first two arguments. If the first is less than or equal to the second, IFLTE returns the value of its third argument. Otherwise it returns the value of its fourth argument.

10.8.2 Terminal set

The terminals or leaves of the trees being evolved by the GP are either pre-calculated compound features (cf. Section 10.7) or constants (see Table 10.2). GProc does not use "ephemeral random constants". Instead all the numeric values used by the evolved expressions are chosen from the fixed initial set. The random constants are drawn from a very non-uniform distribution of both positive and negative values, with about 500 values lying between -10 and +10. This is generated using a "tangent" distribution [10.42]. Random values are uniformly generated in the range $0 \ldots \pi$ and then scaled by multiplying by 10. Duplicates are discarded. Finally the integers 0 to 9 are also included to assist with categorical data. Fig. 10.7 shows the distribution of constants.

10.8.3 GP Genetic Operations and other Parameters

Following [10.44] and others, we use a high mutation rate and a mixture of different mutation operators. To avoid bloat, we also use size fair crossover [10.43] and limit the maximum model size to 63, see Table 10.2.

[1] http://www.daylight.com/dayhtml/smiles/

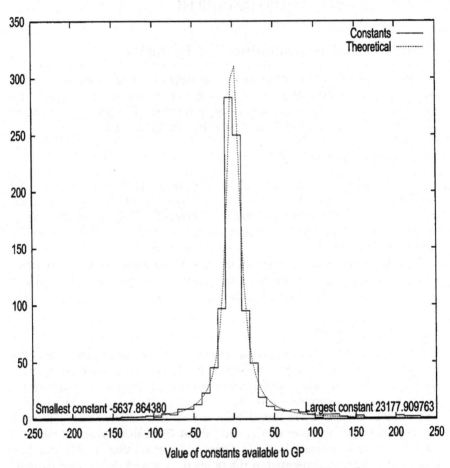

Fig. 10.7. Distribution of fixed constants Theoretical line ($\frac{1}{r\pi} \frac{1}{1+(x/r)^2}$, $r = 10$) is derived from derivative of inverse of tangent function used to randomly chose the constants. Scaling factor r means we expect 50% of constants to lie in $-r \ldots r$, 10% to lie outside $-6.314r \ldots 6.314r$, 1% to lie outside $-63.66r \ldots 63.66r$, 0.1% to lie outside $-636.6r \ldots 636.6r$ etc.

Table 10.2. GP Parameters

Objective:	Evolve a predictive model of human bioavailability
Function set:	MUL ADD DIV SUB IFLTE
Terminal set:	83 features, 0 0 1 2 3 4 5 6 7 8 9 plus 1000 unique random constants
Fitness:	$100,000 \times \left(\frac{1}{2}\text{TP} + \frac{1}{2}(1 - \text{FP})\right) - \sum |a - actual|^2 /$num chemicals. Measured on 321 (59- and 262+) drugs selected for training
Selection:	generational (non elitist), tournament size 7
Wrapper:	Force into range 0..100 (i.e. if $a < 0, a = 0$ if $a > 100, a = 100$) Iff $a > 33$ Predict ok.
Pop Size:	500
Max model size:	63
Initial pop:	Each individual comprises one tree each created by ramped half-and-half (2:6) (half terminals are constants)
Parameters:	50% size fair crossover, crossover fragments $\leq$ 30 [10.43] 50% mutation (point 22.5%, constants 22.5%, shrink 2.5% subtree 2.5%)
Termination:	generation 50

10.8.4 GP Fitness Function

Each individual returns a real number. This is treated as if it was a prediction of the true percentage bioavailability. However first it is truncated to force it to lie in the range $0 \ldots 100$. There are two components of fitness 1) error squared and 2) $\frac{1}{2}\text{TP} + \frac{1}{2}(1 - \text{FP})$ (cf. Section 10.5.1).

Error squared is simply the sum of the squared difference between the (truncated) value returned by the individual and the measured bioavailability across all the training compounds, divided by the number of training compounds.

To calculate the true positive (TP) and false positive (FP) rates, the (truncated) value returned by the tree is compared with the threshold 33. Values ≤ 33 are treated as predicting poor bioavailability (negative) while values above 33 indicate positives. The individual's overall fitness is given by the weighted sum of the two components: $f = 100,000 \times \left(\frac{1}{2}\text{TP} + \frac{1}{2}(1 - \text{FP})\right) + \frac{\sum |error|^2}{\text{number cases}}$. (100,000 was chosen as the weighting factor empirically. A Pareto approach might have been used instead [10.42].)

10.9 Experiments

We conducted two experiments, both of five runs. The first used 321 compounds randomly drawn from the first 481 human data records. The second set used 1342 records randomly chosen from the 2013 records for rats. Apart from the training data, the second set of runs were identical to the first, cf. Table 10.2.

10.9.1 Training on 321 Human Records

The fittest models in generation 50 of the five runs were compared. The model shown in Fig. 10.8 was chosen as the best overall. This was primarily because the difference between its (ROC) performance on the training and 160 human test records was the smallest of the five. A small difference suggests that the model does not over fit the training data and so may generalize to other chemicals. While the test data was used for model selection, the model's performance was later tested on a further 124 holdout records. No significant difference was found. We are reasonable confident that the model's ROC ("605" in Fig. 10.9) is a fair indication of its likely performance on *similar* data.

After model selection, it was tested on the 2013 compounds whose bioavailability in rats was known. Its performance was significantly worse.

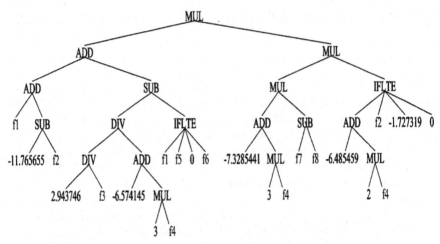

Fig. 10.8. Evolved model of bioavailability created using human training data. The model has been simplified to remove redundant code, without affecting its performance. Final size 55

10.9.2 Training on 1342 Rat Records

Again the five fittest models evolved at the end of the five runs where compared, and again the one (shown in Fig. 10.10) was chosen since it had the smallest difference between training and test performance. Its performance on its training data is noticeably worse than that trained on the human data set. However (see Fig. 10.11) its performance does not fall away when tested on data from another species (i.e. human). That is, our second experiments automatically produced a simpler model with wider applicability than the first but at the cost of lower predictive performance.

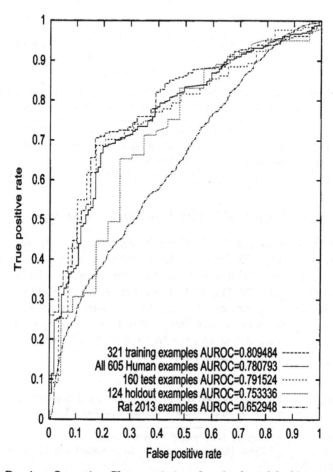

Fig. 10.9. Receiver Operating Characteristics of evolved model of bioavailability created using human training data. (Model is shown in Fig. 10.8. It has been simplified by hand). Performance across all human data includes (20%, 68%) while for the same (68%) true positive rate the classifier only achieves FP=51% across all rat data. There is no statistical significance [10.41] between the area under the curves (AUROC) for the four human datasets. However the difference between AUROCs of the combined human and rat curves is unlikely to be due to chance fluctuations

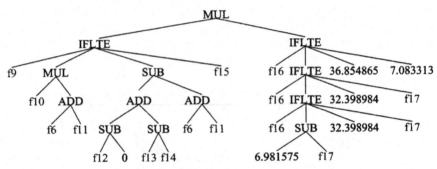

Fig. 10.10. Evolved model of bioavailability created using rat training data. Size 35. Only feature f6 is used by both this model and that shown in Fig. 10.8. One can readily see that the model falls into the product of left and right components. The left evaluated to either f11+f12+f13-f14-f16 or f15 depending if f9≤f10(f6+f11). While the right hand side comes to either a small value (7.083313) or a large (36.854865) value depending only on f16 v. f17. The RHS can be simplified to if f16> 6.981575−f17 then 7.083313 otherwise 36.854865, with only marginal effect

10.9.3 Simplification of Evolved Model of Rat Bioavailability

As Fig. 10.10 makes clear the GP model evolved using the rat data is unnecessarily complex. The first of three stages of simplification was to replace the large nested if statement on the right hand side (RHS) by a single if (actually by (IFLTE f16 (SUB 6.981575 f17) 36.854865 7.083313), cf. Fig. 10.12. In the second stage, the new model was used to seed a new GP run. In order to encourage the evolution of still simpler models, their maximum size was reduced and the fraction of shrink mutation was greatly increased. Details are given in Table 10.3.

Table 10.3. GP Parameters used to simplify model, cf. Table 10.2. (All parameters were as the first set of GP runs on rat training data except where given)

Objective:	Simplify best model evolved using rat bioavailability data
Selection:	Elitist. (Generational and tournament size 7, as before)
Max model size:	41
Initial pop:	100% seeded with model shown in Fig. 10.12. Each new subtree inserted by subtree mutation is created with ramped half-and-half as before but with max depth range (1:2), rather than (2:6).
Parameters:	10% size fair crossover, 90% mutation (point 4.5%, constants 4.5%, shrink 76.5%, subtree 4.5%)

After 50 generations, this run produced a simplified model of similar performance. The RHS was unchanged but the LHS was simplified. In the new model the the first argument of the if (f9) was replaced by f15, two of the features in the second argument were replaced and the large subtree which

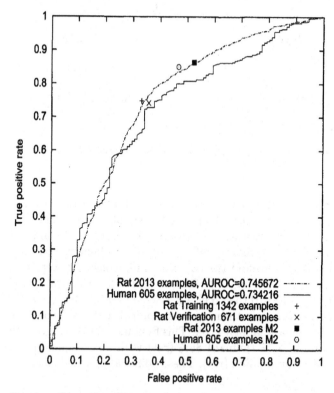

Fig. 10.11. Receiver Operating Characteristics of evolved model of bioavailability created using rat training data (cf. Fig. 10.10). This model achieves a false positive rate of 32% for a true positive rate of 70%. Only a single point on the ROC (shown with +) is used to assess the performance of the classifiers as they are evolved. (The performance of this classifier on the rat validation data is shown with ×.) Hanley's statistical significance test [10.41] shows the difference in the AUROC of the rat and human curves can be explained as chance fluctuations. The two M2 points refer to the simplified model shown in Fig. 10.15

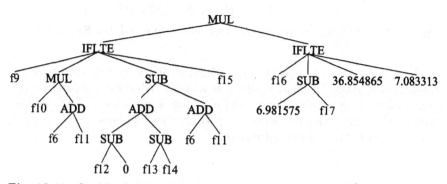

Fig. 10.12. Seed loaded into GP population for automatic simplification. Size 27. (Produced by shrinking RHS of model evolved using rat data, cf. Fig. 10.10)

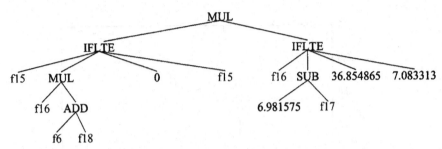

Fig. 10.13. Automatic simplification of seed model (cf. Fig. 10.12) produced at generation 50 by GP. Size 17

had been the third argument of the if was replaced by zero, see Fig. 10.13. In the third stage the model was further simplified by hand.

The model was rewritten to reverse the order of the multiplication and the ifs (making no semantic difference). The subexpression f16 × (f6 + f18) was replaced by zero and the two remaining constants were rounded to the nearest integer. Yielding if(f_1(f15, f16, f17) > 33 then bioavailability is ok. Where f_1 = if f15 ≤ 0 then 0 else if f16 < (7 − f17) then f15 × 37 else f15 × 7, cf. Fig. 10.14. Which in turn can be simplified to if f15 > 0 and f16 < (7 − f17) then predict bioavailability is ok. This final model (M2, Fig. 10.15) has a true positive rate of 85% (human) 88% (rat) with a corresponding false positive rate of 47% (human) 55% (rat). These points are plotted as M2 on Fig. 10.11.

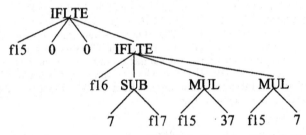

Fig. 10.14. Simplification of compacted model slimed by GP (cf. Fig. 10.13)

Note, apart from re-arranging the if and multiplication operators, each simplification made by hand resulted in a slight reduction in classification performance. In principle, if one could establish a quantifiable trade off between model complexity and accuracy (perhaps based on information theory) these steps could have been automated.

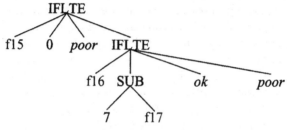

Fig. 10.15. Predicting bioavailability (M2 production version)

10.10 Discussion

The results of the previous section indicate that we can automatically evolve predictive models of bioavailability for chemicals like those for which we already have measurements.

A, possibly simplistic, explanation for the difference between the results when training with human data and rat data is simply the size and nature of the training data. When the volume of data is small its easy to get higher performance in the limited domain. However, in these cases, machine learning models are liable not to extrapolate out of their training domain. Thus the first model works on human data but fails on the rat data. In contrast, Fig. 10.5 (page 222) suggests that the rat data is more diverse (and thus harder to learn) but also covers the space occupied by marketed drugs as a subset. Since modelling the rat data is harder, performance is somewhat reduced but, importantly, it does not fall off on the human data set. We tried to confirm these assumptions using a cluster analysis but this was somewhat inconclusive. Obtaining identical performance on different species with different chemicals is very encouraging.

It is critical that models are able to predict properties of chemicals, when we do not know them. Our experiments tend to reinforce the view that this is possible only where chemicals are like those we have already met. Accuracy will fall away with dissimilar chemicals. Nonetheless in drug discovery we are dealing with a very limited part of the whole of chemical space and so we may hope for useful extrapolation.

The simplicity of the evolved model (Fig. 10.15) suggests that our initial assumption that many mechanisms are involved in ensuring molecules reach the blood stream was too cautious. The simplicity hints that perhaps either there are few dominating mechanisms or that many mechanisms are similar. A second, less encouraging, possibility is that few mechanisms are involved with the observed chemicals because the chemicals we have measurements for, are too similar to each other. That is, a more complex model would be needed to cover the whole chemical space of potential drugs. Also the model suggests that the GSK features are indeed suitable for modelling important drug properties.

Other pharmaceutical uses of data mining might be to highlight which events in patient histories are important and to sift through records to differentiate drug interactions from the normal background prevalence of unrelated or unchanged diseases. Following the success of the Human Genome project, it is anticipated that DNA data (e.g. gene expression levels) will soon lead to the rapid identification of patients with genetic predispositions to both disease and adverse reactions to drug treatments. However, at least at present, DNA chip data presents a difficult data mining problem.

10.11 Conclusion

We have used genetic programming (GP) to automatically create interpretable predictive models of a small number of very complex biological interactions of great interest to medicinal and computational chemists who search for new drug treatments. Particularly, we have found a simple predictive model of human oral bioavailability (Fig. 10.15). While the models are not (and probably can never be) 100% accurate, they use readily available data and so can be used to guide the choice of which molecules to forward to the next (more expensive) stage in the drug selection process. Notably the models can make *in silico* predictions about "virtual" chemicals, e.g. to decide if they are to be synthesized.

Since the models are simple and presented as mathematical functions, they can be readily ported into other tools, e.g. spreadsheets, database queries and intranet pages. Little more than "cut and paste" may be required. Run time of the GP part of the model (once produced) is unlikely to be an issue.

Acknowledgements

We would like to thank Sandeep Modi and Chris Luscombe.

References

10.1 Livingstone, D. (1995): Data analysis for chemists – applications to QSAR and chemical product design. Oxford University Press
10.2 Van de Waterbeemd, J., Gifford, E. (2003): ADMET in silico modelling: towards prediction paradise? Nature Reviews Drug Discovery, **2**, 192–204
10.3 Lipinski, C. A., Lombardo, F., Dominy, B. W., Feeney, P. J. (1997): Experimental and computational approaches to estimate solubility and permeability in drug discovery and development settings. Advanced Drug Delivery Reviews, **23**, 3–25
10.4 Ekins, A., Waller, C. L., Swaan, P. W., Cruciani, G., Wrighton, S. A., Wikel, J. M (2000): Progress in predicting human ADME parameters in silico. Journal of Pharmacological and Toxicological Methods, **44**, 251–272
10.5 Journal of chemical information and computational sciences.

10.6 Podlogar, B. L., Muegge, I. (2001): "Holistic" in silico methods to estimate the systemic and CNS bioavailabilities of potential chemotherapeutic agents. Current Topics in Medicinal Chemistry, 1, 257–275

10.7 Jones, G. (1998): Genetic and evolutionary algorithms. In Encyclopedia of Computational Chemistry. John Wiley & Sons, Ltd

10.8 Sheridan, R. P., SanFeliciano, S. G., Kearsley, S. K. (2000): Designing targeted libraries with genetic algorithms. Journal of Molecular Graphics and Modelling, 18, 320–334

10.9 Nicolotti, O., Gillet, V. J., Fleming, P. J., Green, D. V. S. (2002): Multi-objective optimization in quantitative structure-activity relationships: Deriving accurate and interpretable QSARs. Journal of Medicinal Chemistry, 45, 5069–5080

10.10 Kotanchek, M., Kordon, A., Smits, G., Castillo, F., Pell, R., Seasholtz, M. B., Chiang, L., Margl, P., Mercure, P. K., Kalos, A. (2002): Evolutionary computing in Dow Chemical. In Lawrence "Dave" Davis and Rajkumar Roy, editors, GECCO-2002 Presentations in the Evolutionary Computation in Industry Track, 101–110, New York, New York

10.11 Langdon, W. B., Barrett, S. J., Buxton, B. F. (2002): Genetic programming for combining neural networks for drug discovery. In Roy, R., Köppen, M., Ovaska, S., Furuhashi, T., Hoffmann, F., editors, Soft Computing and Industry Recent Applications, 597–608. Springer-Verlag, 10–24

10.12 Langdon, W. B., Barrett, S. J., Buxton, B. F. (2002): Combining decision trees and neural networks for drug discovery. In Foster, J. A., Lutton, E., Miller, J., Ryan, C., Tettamanzi, A. G. B., editors, Genetic Programming, Proceedings of the 5th European Conference, EuroGP 2002, 2278 of LNCS, 60–70, Springer-Verlag

10.13 Langdon, W. B., Barrett, S. J., Buxton, B. F. (2003): Comparison of adaboost and genetic programming for combining neural networks for drug discovery. In Raidl, G., Cagnoni, S., Cardalda, J. J., Corne, D. W., Gottlieb, J., Guillot, A., Hart, E., Johnson, C. J., Marchiori, E., Meyer, J. A., Middendorf, M., editors, Applications of Evolutionary Computing, EvoWorkshops2003: EvoBIO, EvoCOP, EvoIASP, EvoMUSART, EvoROB, EvoSTIM, 2611 of LNCS, 87–98, Springer-Verlag

10.14 Langdon, W. B., Barrett, S. J., Buxton, B. F. (2003): Predicting biochemical interactions – human P450 2D6 enzyme inhibition. In Proceedings of the 2003 Congress on Evolutionary Computation CEC2003, IEEE Press

10.15 Bains, W., Gilbert, R., Sviridenko, L., Gascon, J. M., Scoffin, R., Birchall, K., Harvey, I., Caldwell, J. (2002): Evolutionary computational methods to predict oral bioavailability QSPRs. Current Opinion in Drug Discovery and Development, 5, 44–51

10.16 Agorama, B., Woltosza, W. S., Bolger, M. B. (2001): Predicting the impact of physiological and biochemical processes on oral drug bioavailability. Advanced Drug Delivery Reviews, 50(Supplement 1), S41–S67,

10.17 Koza, J. R. (1992): Genetic programming: on the programming of computers by means of natural selection. MIT Press

10.18 Banzhaf, W., Nordin, P., Keller, R. E., Francone F. D. (1998): Genetic programming – an introduction; on the automatic evolution of computer programs and its applications. Morgan Kaufmann

10.19 Banzhaf, W., Foster, J. A. (editors.) (2002): Biological applications of evolutionary computation (BioGEC 2002), AAAI

10.20 Banzhaf, W., Foster, J. A. (editors.) (2003): Biological applications of evolutionary computation (BioGEC 2003), AAAI

10.21 Marchiori, E., Corne, D. W., (editors.) (2003): EvoBIO, the first european workshop on evolutionary bioinformatics, volume **2611** of LNCS, Springer-Verlag

10.22 Kell, D. (2002): Defence against the flood. Bioinformatics World, 16–18

10.23 Moore, J. H., Parker, J. S., Olsen, N. J., Aune, T. M. (2002): Symbolic discriminant analysis of microarray data in automimmune disease. Genetic Epidemiology, **23**, 57–69

10.24 Iba, H., Sakamoto, E. (2002): Inference of differential equation models by genetic programming. In Langdon, W. B., Cantú-Paz, E., Mathias, K., Roy, R., Davis, D., Poli, R., Balakrishnan, K., Honavar, V., Rudolph, G., Wegener, J., Bull, L., Potter, M. A., Schultz, A. C., Miller, J. F., Burke, E., Jonoska, N., editors, GECCO 2002: Proceedings of the Genetic and Evolutionary Computation Conference, 788–795, Morgan Kaufmann Publishers

10.25 Koza, J. R., Mydlowec, W., Lanza, G., Yu, J., Keane, M. A. (2000): Reverse engineering of metabolic pathways from observed data by means of genetic programming. In First International Conference on Systems Biology (ICSB)

10.26 Langdon, W. B., Buxton, B. F. (2001): Evolving receiver operating characteristics for data fusion. In Miller, J. F., Tomassini, M., Lanzi, P. L., Ryan, C., Tettamanzi, A. G. B., Langdon, W. B., editors, Genetic Programming, Proceedings of EuroGP'2001, volume **2038** of LNCS, 87–96, Springer-Verlag

10.27 Langdon, W. B., Buxton, B. F. (2001): Genetic programming for combining classifiers. In Spector, L., Goodman, E. D., Wu, A., Langdon, W. B., Voigt, H. M., Gen, M., Sen, S., Dorigo, M., Pezeshk, S., Garzon, M. H., Burke, E., editors, Proceedings of the Genetic and Evolutionary Computation Conference (GECCO-2001), 66–73, Morgan Kaufmann

10.28 Langdon, W. B., Buxton, B. F. (2001): Genetic programming for improved receiver operating characteristics. In Josef Kittler and Fabio Roli, editors, Second International Conference on Multiple Classifier System, volume **2096** of LNCS, 68–77, Springer Verlag

10.29 Langdon, W. B., Buxton, B. F. (2004): Genetic programming for mining DNA chip data from cancer patients. Genetic Programming and Evolvable Machines

10.30 Yoshida, F., Topliss, J. G. (2000): QSAR model for drug human oral bioavailability. Journal of Medicinal Chemistry, **43**, 2575–2585

10.31 Andrews, C. W., Bennett, L., Yu, L. X. (2000): Predicting human oral bioavailability of a compound: Development of a novel quantitative structure-bioavailability relationship. Pharmaceutical Research, **17**, 639–644

10.32 Pintore, M., van de Waterbeemd, H., Piclin, N., Chretien, J. R. (2003): Prediction of oral bioavailability by adaptive fuzzy partitioning. European Journal of Medicinal Chemistry, **38**, 427–431. XVIIth International Symposium on Medicinal Chemistry

10.33 Veber, D. F., Johnson, S. R., Cheng, H. Y., Smith, B. R., Ward, K. W., Kopple, K. D.(2002): Molecular properties that influence the oral bioavailability of drug candidates. Journal of Medicinal Chemistry, **45**, 2615–2623

10.34 Mandagere, A. K., Thompson, T. N., Hwang, K. K. (2002); Graphical model for estimating oral bioavailability of drugs in humans and other species from their Caco-2 permeability and in vitro liver enzyme metabolic stability rates. Journal of Medicinal Chemistry, **45**, 304–311

10.35 Goldberg, D. E. (1989): Genetic algorithms in search optimization and machine learning. Addison-Wesley

10.36 Langdon, W. B., Soule, T., Poli, R., Foster, J. A. (1999): The evolution of size and shape. In Lee Spector, William B. Langdon, Una-May O'Reilly, and

Peter J. Angeline, editors, Advances in Genetic Programming, **3**, 163–190. MIT Press

10.37 Swets, J. A., Dawes, R. M., Monahan, J. (2000): Better decisions through science. Scientific American, **283**, 70–75

10.38 Scott, M.J.J., Niranjan, M., Prager, R. W. (1998): Realisable classifiers: improving operating performance on variable cost problems. In Lewis, P. H., Nixon, M. S., editors, Proceedings of the Ninth British Machine Vision Conference, **1**, 304–315, University of Southampton, UK

10.39 Provost, F., Fawcett, T. (2001): Robust classification for imprecise environments. Machine Learning, **42**, 203–231

10.40 Yusoff, Y., Kittler, J., Christmas., W. (1998): Combining multiple experts for classifying shot changes in video sequences. In IEEE International Conference on Multimedia Computing and Systems, **2**, 700–704, Florence, Italy

10.41 Hanley, J. A., McNeil, B. J. (1982): The meaning and use of the area under a receiver operating characteristic (ROC) curve. Radiology, **143**, 29–36

10.42 Langdon, W. B. (1998): Genetic programming and data structures: genetic programming + data structures = automatic programming!, **1**, Genetic Programming. Kluwer, Boston

10.43 Langdon, W. B. (2000): Size fair and homologous tree genetic programming crossovers. Genetic Programming and Evolvable Machines, **1**, 95–119

10.44 Angeline, P. J. (1998): Multiple interacting programs: A representation for evolving complex behaviors. Cybernetics and Systems, **29**, 779–806

11. Microarray Data Mining with Evolutionary Computation

Gary B. Fogel

Natural Selection, Inc.
3333 N. Torrey Pines Ct., Suite 200
La Jolla, California 92037
USA
gfogel@natural-selection.com

11.1 Introduction

To say that computers have played an important role in the development of
modern molecular biology is a drastic understatement. It is difficult to imag-
ine modern biology without computational assistance, especially with respect
to specialized algorithms for data analysis and interpretation. With better
computation comes answers to a variety of new and interesting complex biological problems that were considered unreachable only a decade previously.
We are fortunate to live in a time where questions that were simplified previously for the sake of convenient mathematics can now be solved without
resorting to assumptions and differential equations and where tomorrow's
questions lie in a better interpretation of large data sets with computers,
whether those computers are supercomputers, farms of Linux processors, or
simple laptop computers. It is an interesting time to be in science, especially
when one considers that the supercomputers of today will be the laptops of
tomorrow.

Many biological problems can be restated in the form of an optimization
problem to provide a reasonable approximation of a globally optimal solution
(e.g., protein folding, drug docking, etc.). This is also the case when developing models for the examination of large data sets such that trends in the data
can be extracted for a better understanding of biological function. Such an
understanding leads directly to better remediation through biochemical action (i.e., drugs, chemotherapy, etc.) in the case of cancer and other diseases.
Thus, a main effort in computational biology or bioinformatics (the interdisciplinary fields that bring together biology, computer science, mathematics,
statistics, and information theory to analysis biological data for interpretation
and prediction) is the development of data mining tools that can assist the
biologist in making key decisions that will reflect in more desirable courses of
medical action and/or better future experimentation. Methods of optimization are central to the success of this effort. The focus of this chapter is the
use of machine learning algorithms including artificial neural networks and
evolutionary computation for the analysis of gene expression data

11.2 Microarrays

In the eight years since its invention [11.14, 11.31, 11.39], DNA microarray technology has rapidly gained attention as a method for the study of global gene expression in cells and tissues, in different experimental conditions over time to assist in the mapping of genotype to phenotype. These approaches have proved to be versatile, with direct application in cancer diagnosis and treatment, drug discovery, and sequencing [11.25, 11.52]. For instance, microarray technologies are currently being applied to early diagnosis of cancer through a better understanding of how certain genes and their proteins contribute to the onset of disease and how they influence patient response to therapy [11.43]. It has been estimated that nearly every molecular biology laboratory in the world will have access in the near future to high-throughput functional-genomic microarray technology and be involved in capturing data sets consisting of tens of thousands of gene expression data points per sample measured [11.3].

DNA microarrays are generally consist of a 2-dimensional surface of samples each containing short strings of nucleic acids (oligonucleotides) per position. There may be as many as 100 sample locations per cm^2 or as many as 25,000 samples per cm^2 where each sample contains thousands of oligonucleotides. Microarrays are produced by robotics in the lab (in the case of spotted arrays) or can be purchased pre-prepared (e.g., GeneChip® brand arrays, Affymetrix, Santa Clara, CA, and others). With microarrays, each oligonucleotide on the chip is matched with an almost identical copy, differing by a single nucleotide substitution. This difference results in a method of internal control for hybridization specificity and determination of nonspecific binding to that sample location. Microarrays can be used for the interrogation of large numbers of genes simultaneously to determine which genes were expressed or not. For genes that are expressed, microarrays can be used to resolve the level of expression. Thus, using these methods it is possible to examine whole genome expressions in different environmental conditions to determine which gene networks are correlated with phenotypic response. It is also possible to examine cancer cells to determine their different expression patterns with respect to "normal" cells and use this as a means of cell classification and/or prediction of outcome. Given that some cancers are indistinguishable by histology, tumor-specific molecular markers are valuable in better diagnosis and assignment of proper chemotherapy treatment [11.1].

The data resulting from microarray analysis exists in the form of a matrix containing a number of genes and their level of expression in different cell types or in a single genome over time. Microarray data is known to have a low signal-to-noise ratio and the reliability of single samples has been questioned [11.13, 11.55]. Therefore many researchers repeat samples for each physiological environment and probabilistic methods of interpretation have been described [11.47]. Data mining methods are then required to either use the gene expression patterns to classify or assign a particular expression pattern as one of a number of known cell types, or to cluster the available expression data into groups of genes that share similar expression values and might be coregulated.

11.3 Supervised and Unsupervised Methods for Microarray Analysis

Supervised methods of machine learning make use of a training set where gene expressions and cell class are known. These training sets can be derived from experimentation on, for instance, cells previously identified to be cancers by histological analysis such that the "truth" condition is known *a priori*. A main objective of supervised methods is the result of a model that relates expression to cell class in a most parsimonious manner. Unsupervised approaches can be used to define clusters in the expression patterns and their relation to cell type that might result in a better understanding of coregulation where the cluster number might not be known *a priori*. Both approaches are useful under different circumstances and with different desired result.

11.3.1 Supervised Methods of Class Assignment

Detailed reviews of the application of supervised methods to microarray-based cancer diagnosis are available elsewhere n the literature [11.49, 11.50, 11.57]. For the sake of comparison to the evolutionary methods described later in this chapter, supervised methods of ANNs [11.28], support vector machines (SVMs) [11.19], and correlation-based classification methods [11.23] will be briefly reviewed here with particular interest in ANNs.

Correlation-based classification methods work well in situations where there is a linear relationship between the gene expressions and output condition. ANNs and SVMs have the ability to map the interactions of gene collections to a prediction of output (i.e., cell type, expression over time, etc.) in a nonlinear fashion, and as a result, these latter methods are likely to be more robust and generating useful predictive models as it is widely expected that genes in a cell interact in a nonlinear fashion in the mapping of genotype to phenotype for a given environment.

Supervised learning with ANNs follows a process of training, validation, and/or testing over a set of exemplars. During training and validation, examples of known correlations between gene expression patterns and desired output (cancer cell class) are examined. The methods and duration of training are critical to overall performance, for instance, the affects of genomic data sampling and its effect on classification performance assessment were reviewed in [11.8] and the predictive quality assessment was found to be dependent on the data size, sampling techniques and the number of train-test experiments. However the resulting ANN serves as a predictive model relating relative importance of each gene expression input to the desired output in a nonlinear fashion [11.46].

Demonstration of the robustness of ANNs in class assignment [11.28], provided evidence to suggest that they could accurately classify data on the expression profiles of 96 transcripts representing 80 known genes and 13 anonymous expressed sequence tags characterizing four distinct subtypes of small round blue cell tumors (SRBCT). Small round blue cell tumors (so named because of their morphological similarity) include neuroblastoma, rhabdomyosarcoma, non-Hodgkin lymphoma, and the Ewing family of tumors. For these cancers there is a wide assortment of required treatments making the problem of appropriate class assignment very important. 88 experiments (63 training and 25 testing) for 6567 genes were filtered for a minimum level of expression (resulting in a set of 2308 genes remaining). Principal component analysis (PCA) was used to reduce the inputs to the 10 most dominant for classification into 1 of 4 cell types. 3750 artificial neural networks trained via backpropagation were then used to assay performance. The average number of misclassified samples for all 3750 models was plotted against increasing number of genes used as input for the neural network. The misclassification was minimized to zero when a set of 96 highest ranked genes was used. When using these 96 genes (and 10 most dominant inputs for each) the neural network was clearly capable of distinguishing between the 4 cancer cell types with 100% accuracy. 41 of the 96 genes were not previously identified as having any association with these 4 cancer types.

The methods detailed in [11.12, 11.28, 11.42, 11.48, 11.54] have demonstrated the utility of ANNs for cancer class assignment. Backpropagation [11.56] s a common method for neural network training in many of these cases but can lead to premature convergence to locally optimal solutions. Alternative stochastic training methods (such as evolutionary computation (EC)) may lead to improved performance during training and validation through a more exhaustive search over the possible net architectures and weights associated with the importance of each node. This will be described below.

Cellular neural networks [11.6], probabilistic neural networks [11.11], bayesian neural networks [11.38], fuzzy methods [11.5, 11.7], and have been applied to the microarray class assignment problems. Training via backpropagation for optimized weight assignment has also been applied with some success. For instance, O'Neill and Song [11.45] utilized neural networks trained via backpropagation to predict the long-term survival of individual patients with 100% accuracy. This research reduced the number of genes associated with diffuse large B-cell lymphoma classification to less than three dozen.

Backpropagation. Multilayer feed forward perceptrons are a common ANN architecture and training these networks can be accomplished through a backpropagation algorithm that implements a gradient search over the error response surface for the set of weights that minimizes the sum of the squared error between the actual and target values.

Backpropagation may be incapable of discovering an appropriate set of weights even if the ANN topology provided sufficient complexity because the procedure provides guaranteed convergence, but only to a locally optimal solution. The researcher is then left to use ad hoc tricks to avoid this premature convergence including the addition of noise to the exemplars, collecting new data and retrain, or adding degrees of freedom to the network by increasing the number of nodes and/or connections.

The last approach presents problems to the designer of the network, for any function can map any measurable domain to its corresponding range if given sufficient degrees of freedom. Unfortunately, such overfit functions generally provide very poor performance during analysis on the test and/or validation sets. The problems of local convergence with the backpropagation algorithm indicate the desirability of training with stochastic optimization methods such as simulated evolution that can provide convergence to globally optimal solutions.

11.4 Evolutionary Computation

Simulated evolution presents a valuable optimization method in the context of bioinformatics. EC includes methods of genetic algorithms (GA), evolutionary programming (EP), evolution strategies (ES), and other derivatives of these techniques such as genetic programming (GP) and classifier systems [11.10]. These evolutionary methods are broadly similar in that all evolutionary algorithms (EAs) maintain a population of contending solutions for a given problem [11.16, 11.17]. Each solution in the population is then scored with respect to a figure or merit or "fitness" that describes its worth. A process of selection is used to remove solutions of low fitness from the population. The solutions that remain after selection are used as "parents" to generate new "offspring" solutions with random variation until the original population size is re-established. This process of selection and variation is iterated until a termination criterion (e.g., a previously specified number of "generations," or until the population has converged on a solution of adequate worth, or until sufficient time no longer remains for the user, etc.) has been exhausted.

The methods of EC tend to differ in terms of the favorite choices of solution representation, the manner and rate in which variation is applied, and the forms of selection that are used. These choices are largely problem dependent, thus any one representation, variation, or selection setting may not be successful over all biological problems, requiring user knowledge to determine which settings to vary for improved performance. These algorithms have been shown to generate robust performance across a wide range of function optimization problems [11.9, 11.16, 11.40].

11.4.1 Evolutionary Computation in Bioinformatics

EC has been applied to problems in the area of bioinformatics over 15 years including drug design [11.20], protein folding [11.24], structure discovery [11.18], prediction of activity from drug structure [11.30] and a wide assortment of problems of pattern recognition in biomedicine, EC can quickly achieve near-optimal solutions to these complex problems even when the state space of possible solutions is too large to be searched exhaustively in real time.

In the context of pattern recognition, EC can be used to evolve optimized data representations (i.e., neural network, finite state machine, etc.) that can be used as a predictive model of a biological process. This main focus of this chapter is the application of evolved neural networks (ENNs) in the area of microarray analysis. Stochastic search techniques such as simulated annealing and evolution consistently outperform backpropagation yet can be executed more rapidly on an appropriately configured parallel processing computer. After sufficient computational effort, the most successful network can be put into practice. For additional information on designing neural networks through simulated evolution, the reader is directed to [11.30].

11.5 Applications of Evolutionary Computation for Class Prediction and Expression Clustering

Hwang et al. [11.29] presented a method for the use of EAs to optimize the structure and weights of radial basis function (RBF) neural networks for the purpose of optimized classification of the Golub et al. dataset [11.23]. Experiments were performed to varying the number of input genes and the number of hidden nodes. This method was able to achieve perfect classification on the training examples but classification accuracy varied on the test examples depending on the parameters and structure of the network being optimized. Similar work was reported by Keedwell [11.26] and Keedwell and Narayanan [11.27] through the development of a novel GA that could identify new gene expression combinations and test for their worth in terms of class prediction accuracy through an ANN. The results from this method were encouraging and suggest that methods of hybridization are important considerations in future method development.

Li et al. [11.33]-[11.37] presented methods that coupled GAs with a non-parametric pattern recognition k-nearest neighbor (KNN) method for the purpose of gene expression classification into one of two classes (tumor and normal). The GA was used to select subsets of 50 gene expressions that could be used to potentially distinguish between tumor and normal tissue samples. The dimensionality of 50 was chosen as a compromise between the rate at which samples become more dissimilar when more genes are compared, the increased noise in the system as a result of larger dimensions, and the computational expense at larger dimensions. For each of the 50 genes, the class memberships of the three nearest neighbors were compared. If all samples had the same class membership, a value of 1 was recorded for that clustering. This process was repeated for all 50 genes and a sum of the cluster values was used as a representation of overall worth. This method was tested on the data sets of Alon et al. [11.2], Alizadeh et al. [11.1] and Golub et al. [11.23] through a series of training and testing examples. The results demonstrated that the GA/KNN method was capable of discovering subsets of genes that were well correlated with sample classification.

Ooi and Tan [11.44] developed a similar hybridized approach for this problem. A GA-based method of gene selection was used to determine the optimal number and set of gene expressions that maximized classification using a maximum likelihood (MLHD) method. To develop an MLHD classifier, the data was separated into training and testing sets and a discriminant function (maximum likelihood) was used to assign samples to the class with the highest conditional probability. The GA then was used to search for subsets of genes that satisfied this discriminant function in the training set, where the worth of the overall approach was examined by prediction on a test set corresponding to nine tumor types [11.51]. The approach was able to generate higher classification accuracies than other published predictive methods on the same multi-class test data set. Ooi and Tan [11.44] outlined several reasons why their approach might outperform the GA/KNN method of Li et al. [11.33]-[11.37] including the decreased sensitivity of the k-nearest neighbor approach with increase in dimensionality, the large memory requirements for storing all the training data, and an approach to post process the n top ranked genes from the 10000 near optimal sets identified by that method based on their frequency of occurrence in the optimal sets. Ooi and Tan [11.44] offered the suggestion that this might work well in the context of binary class predictions, but may not offer the same worth in the case of larger numbers of different classes. These claims remain to be examined.

Deutsch [11.15] utilized a method of "genetic evolution of sub-sets of expressed sequences (GESSES)" to find highly relevant genes by considering different ensembles of predictors and evolving this population with the addition and/or deletion of genes until optimal predictive performance was achieved. The approach builds progressively larger gene clusters. The method was tested on the small round blue cell tumor data of Khan et al. [11.28] and the top 100 gene clusters were used to classify the 4 cell types with 100% accuracy on the test set. This is particularly interesting in light of the fact that the best linear predictors presented in [11.28] utilized 96 gene expressions as inputs whereas the top predictors from Deutsch [11.15] only required 12±2, a much more parsimonious result. The same method was equally successful when tested on a data set of diffuse large B-cell lymphomas [11.53] and outperformed variants of support vector machines and standard KNN approaches to distinguish the 14 types of tumors offered in [11.46] but did not have a high classification accuracy on the leukemia data offered in Golub et al.

[11.23]. This latter result is congruent with a lack of convergence to near perfect predictors on this data observed elsewhere in the literature [11.19, 11.37]. The GESSES method appears to be one of the more extensively developed and studied applications of EA to the problem of class assignment from gene expression.

Ando and Iba [11.4] presented a method for cancer cell classification from gene expression data based on artificial immune systems (AIS). AIS can be considered as an evolutionary model of the immune system as a process of recognition for the purpose of classification. Just as the natural immune system learns to recognize the patterns of antigens, AIS can be developed for pattern recognition tasks. When applied to cancer cell classification, this process was able to discover rules to correctly classify all training and test examples, providing interesting value in terms of a biological understanding of the process.

A description of GP for the classification of yeast gene expression data into functional class has also been offered [11.21, 11.22]. A training set of known genes from each functional class was used to evolve a useful representation from a population of random initial classifier rules. The resulting models had near perfect accuracy on the test set and could be used to discover additional details about the overall expression patterns that correlated well with data in the literature.

The worth of any stochastic optimization process should be measured by repeated sampling with cross validation. As pointed out in [11.41] this is not yet standard in the field. Cross validation consistency has been reviewed [11.41] specifically with regards to GP for mining gene expression patterns in DNA microarray data. This paper is an important contribution towards a refined statistical approach of performance evaluation.

Lee et al. [11.32] provided a means to optimize clusters of gene expressions using ES. For this purpose, the data was represented by a completely connected weighted-graph with similarity measure where the vertex of the graph corresponds to each object to be clustered and the edges represent the similarity between objects. Similarities measures were stored in a pairwise matrix. A successive bipartioning method was used to generate a hierarchical tree. For this purpose, incisional hyperplanes were inserted into the data to decompose the graph into two parts. A matrix incision index was used to score the worth of different incisional hyperplanes with respect to the similarity within clusters and the difference between clusters. An ES was used to optimize the placement of the incisional hyperplanes to maximize the matrix incision index value. This method was able to capture the clustering of the Golub et al. [11.23] data with lower error than bottom-up hierarchical clustering, k-means clustering, or SOMs. With this method, different clustering indices and similarity measures can easily be tried, something which is difficult to do with algorithm-dependent methods such as k-means and SOMs.

11.6 Conclusions

The application of evolutionary algorithms to data mining in the area of
microarray analysis is rapidly gaining interest in the community. The large
number of gene expressions coupled with analysis over a time course, provides
an immense space of possible relations. Some small portion of this space contains information that is of extreme value to modern biomedicine in terms of
proper diagnosis and treatment of many diseases. Classical methods of optimization tend to provide solutions that are of limited worth simply because
of their typical entrapment in locally optimal solutions. Escape from these
local optima cannot be guaranteed with use of evolutionary algorithms, but
is a possibility, and the literature above demonstrates that increased performance is possible. In light of the overabundance of expression data, the
application of methods of simulated evolution towards the development of
better predictive models holds a promising future.

References

11.1 Alizadeh, A. A., Eisen, M. B., Davis, R. E., Ma, C., Lossos, I. S., Rosenwald,
A., Boldrick, J. C., Sabet, H. et al. (2000): Distinct type of diffuse large B-cell
lymphoma identified by gene expression profiling. Nature. 403, 503-511

11.2 Alon, U., Barkai, N., Notterman, D. A., Gish, K., Ybarra, S., Mack, D., Levine,
A. J. (1999): Broad patterns of gene expression revealed by clustering analysis
of tumor and normal colon tissues probed by oligonucleotide arrays. Proc. Natl.
Acad. Sci. USA. 96, 6745

11.3 Anderle, P., Duval, M., Draghici, S., Kuklin, A., Littlejohn, T. G., Medrano,
J. F., Vilanova, D., Roberts, M.A. (2003): Gene expression databases and data
mining. BioTechniques, 34, S36-S44

11.4 Ando, S., Iba, H. (2003): Artificial immune system for classification of cancer.
In: Applications of Evolutionary Computing: EvoWorkshops 2003: EvoBIO,
EvoCOP, EvoIASP, EvoMUSART, EvoROB, and EvoSTIM, Essex, UK, April
14-16, 2003. Springer-Verlag, Heidelberg. 1-10

11.5 Ando, T., Hanai, T., Honda, H., Kobayashi, T. (2001): Prognostic prediction
of lymphoma by gene expression profiling using FNN. Genome Informatics 12,
247-248

11.6 Arena, P., Bucolo M, Fortuna L, Occhipinti, L. (2002): Cellular neural networks for real-time DNA microarray analysis. IEEE Eng Med Biol Mag. 21,
17-25

11.7 Azuaje, F. (2001): A computational neural approach to support the discovery
of gene function and classes of cancer. IEEE Trans Biomed Eng. 48, 332-9

11.8 Azuaje, F. (2003): Genomic data sampling and its effect on classification performance assessment. BMC Bioinformatics. 4, 5

11.9 Bäck, T. (1996): Evolutionary algorithms in theory and practice. Oxford University Press, New York

11.10 Bäck, T., Fogel, D. B., and Michalewicz, Z. (eds.) (1997): Handbook on
evolutionary computation, Oxford University Press, New York

11.11 Berrar, D. P., Downes, C. S., Dubitzky, W. (2003): Multiclass cancer classification using gene expression profiling and probabilistic neural networks. Pac.
Symp. Biocomput. 5-16

11.12 Bicciato, S., Pandin, M., Didone, G., Di Bello, C. (2003): Pattern identification and classification in gene expression data using an autoassociative neural network model. Biotechnol. Bioeng. **81**, 594-606

11.13 Clare, A., King, R. D. (2002): How well do we understand the clusters found in microarray data? In Silico Biol. **2**, 511-522

11.14 DeRisi, J. L., Iyer, V. R., Brown, P. O. (1997): Exploring the metabolic and genetic control of gene expression on a genome scale. Science. **278**, 680-686

11.15 Deutsch, J. M. (2003): Evolutionary algorithms for finding optimal gene sets in microarray prediction. Bioinformatics **19**, 45-52

11.16 Fogel, D. B. (2000): Evolutionary computation: toward a new philosophy of machine intelligence. IEEE Press. Piscataway, NJ. Second Edition

11.17 Fogel, G. B and Corne, D. W. (eds.) (2003) Evolutionary computation in bioinformatics. Morgan Kaufmann, San Francisco

11.18 Fogel, G. B., Porto, V. W., Weekes, D. G., Fogel, D. B., Griffey, R. H., McNeil, J. A., Lesnik, E., Ecker, D. J., Sampath, R. (2002): Discovery of RNA structural elements using evolutionary computation. Nuc. Acids Res. **30**, 5310-5317

11.19 Furey, T.S., Cristianini, N., Duffy, N., Bednarski, D. W., Schummer, M., Haussler, D. (2000): Support vector machine classification and validation of cancer tissue samples using microarray expression data. Bioinformatics **16**, 906-914

11.20 Gehlhaar, D. K., Verkhivker, G. M., Rejto, P. A., Sherman, C. J., Fogel, D. B., Fogel, L. J., Freer, S. T. (1995): Molecular recognition of the inhibitor AG-1343 by HIV-1 protease: conformationally flexible docking by evolutionary programming. Chem. Biol. **2**, 317-324

11.21 Gilbert, R. J., Goodacre, R., Shann, B., Taylor, J., Rowland, J.J., Kell, D. B. (1998): Genetic programming-based variable selection for high-dimensional data. In Genetic Programming 1998: Proceedings of the Third Annual Conference, (Ed. J.R. Koza, W. Banzhaf, K. Chellapilla, K. Deb, M. Dorigo, D.B. Fogel, M.H. Garzon, D.E. Goldberg, H. Iba, and R.L. Riolo), Morgan Kaufmann, San Francisco. 109-115

11.22 Gilbert, R. J., Rowland, J. J., Kell, D. B. (2002): Genomic computing: explanatory modeling for functional genomics. In Proceedings of the Genetic and Evolutionary Computation Conference. (Eds. D. Whitley, D. Goldberg, E. Cantu-Paz), Morgan Kaufmann, San Francisco, 551-557

11.23 Golub, T. R., Slonim, D. K., Tamayo, P., Huard, C., Gaasenbeek, M., Mesirov, J. P., Coller, H., Loh, M. L. et al. (1999) Molecular classification of cancer: class discovery and class prediction by gene expression monitoring. Science. **286**, 531-537

11.24 Greenwood, G. W. and Shin, J. M. (2003) On the evolutionary search for solutions to the protein folding problem, In Evolutionary Computation in Bioinformatics (G.B. Fogel and D.W. Corne eds.), Morgan Kaufmann Pub., San Francisco, 115-136

11.25 He, Y. D., Friend, S. H. (2001): Microarrays-the 21st century diving rod? Nat. Med. **7**, 658-659

11.26 Keedwell, W. (2003): Knowledge discovery from gene expression data using neural-genetic models. Doctoral Dissertation, University of Exeter

11.27 Keedwell, E. and Narayanan, A. (2003): Genetic algorithms for gene expression analysis. In: Applications of Evolutionary Computing: EvoWorkshops 2003: EvoBIO, EvoCOP, EvoIASP, EvoMUSART, EvoROB, and EvoSTIM, Essex, UK, April 14-16, 2003. Springer-Verlag, Heidelberg, 76-86

11.28 Khan, J., Wei, J. S., Ringner, M., Saal, L. H., Ladanyi, M., Westermann, F., Berthold, F., Schwab, M. Antonescu, C. R., Peterson, C., Meltzer, P.S.

(2001) Classification and diagnostic prediction of cancers using gene expression profiling and artificial neural networks. Nat. Med. **7**, 673-679

11.29 Hwang, H. B., Cho, D. Y., Park, S. W., Kim, S. D., Zhang, B. T. (2000): Applying machine learning techniques to analysis of gene expression data: Cancer diagnosis. In: Methods of Microarray Data Analysis: Papers from CAMDA'00 (eds. S.M. Lin and K.F. Johnson), Kluwer Academic Publishers

11.30 Landavazo, D. G., Fogel, G. B., Fogel, D. B. (2002): Quantitative structure-activity relationships by evolved neural networks for the inhibition of dihydrofolate reductase by pyrimidines. BioSystems, **65**, 37-47

11.31 Lashkari, D. A., DeRisi, J. L., McCusker, J. H., Namath, A. F., Gentile, C., Hwang, S. Y., Brown, P. O., Davis, R. W. (1997): Yeast microarrays for genome wide parallel genetic and gene expression analysis. Proc. Natl. Acad. Sci. USA **94**, 13057-13062

11.32 Lee, K., Kim, J. H., Chung, T. S., Moon, B. S., Lee, H., Kohane, I. S. (2001): Evolution strategy applied to global optimization of clusters in gene expression data of DNA microarrays. In Proceedings of the Congress on Evolutionary Computation 2001, IEEE Press, Piscataway, NJ, 845-850

11.33 Li, L., Darden, T. A., Weinberg, C. R., Levine, A. J., Pedersen, L. G. (2001a): Gene assessment and sample classification for gene expression data using a genetic algorithm/k-nearest neighbor method. Comb. Chem. High Through. Screen. **4**, 727-739

11.34 Li, L., Pedersen, L. G., Darden, T. A., Weinberg, C. R. (2001b): Computational analysis for leukemia microarray expression data using the GA/KNN method. In: Methods of Microarray Data Analysis: Papers from CAMDA'00 (eds. S.M. Lin and K.F. Johnson), Kluwer Academic Publishers

11.35 Li, L., Pedersen, L. G., Darden, T. A., Weinberg, C. R. (2001c): Class prediction and discovery based on gene expression data. Presented at the Atlantic Symposium on Computational Biology, Genome Information Systems & Technology 2001 (CBGI)

11.36 Li, L., Weinberg, C. R., Darden, T. A., Pedersen, L. G. (2001d): Gene selection for sample classification based on gene expression data: study of sensitivity to choice of parameters of the GA/KNN method. Bioinformatics. **17**, 1131-1142

11.37 Li, W., Yang, Y. (2002): How many genes are needed for a discriminant microarray data analysis? In Methods of Microarray Data Analysis. Kluwer Academic, 137-150

11.38 Liang, Y., George, E. O., Kelemen, A. (2002): Bayesian neural network for microarray data. In Proceedings of the IEEE International Joint Conference on Neural Networks, 193-197

11.39 Lipshultz, R. J., Fodor, S. P., Gingeras, T. R., Lockhart, D. J. (1999): High density synthetic oligonucleotide arrays. Nat. Genet. **21**, 20-24

11.40 Michalewicz, Z. (1996): Genetic algorithms + data structures = evolution programs. Springer, New York, Third Edition

11.41 Moore, J. H. (2003): Cross validation consistency for the assessment of genetic programming results in microarray studies. In: Applications of Evolutionary Computing: EvoWorkshops 2003: EvoBIO, EvoCOP, EvoIASP, Evo-MUSART, EvoROB, and EvoSTIM, Essex, UK, April 14-16, 2003. Springer-Verlag, Heidelberg, 99-106

11.42 Nikkila J, Toronen P, Kaski S, Venna J, Castren E, Wong G. (2003): Analysis and visualization of gene expression data using self-organizing maps. Neural Netw. **15**, 953-66

11.43 Ochs, M. F., Godwin, A. K. (2003): Microarrays in cancer: research and applications. BioTechniques. **34**, S4-S15

11.44 Ooh, C. H., Tan, P. (2003): Genetic algorithms applied to multi-class prediction for the analysis of gene expression data. Bioinformatics **19**, 37-44

11.45 O'Neill, M. C., Song, L. (2003): Neural network analysis of lymphoma microarray data: prognosis and diagnosis near-perfect. BMC Bioinformatics **4**, 13

11.46 Ramaswamy, S., Tamayo, P., Rifkin, R., Mukherjee, S., Yeang, C. H., Angelo, M., Ladd, C., Reich, M. et al. (2001) Multiclass cancer diagnosis using tumor gene expression signatures. Proc. Natl. Acad. Sci. USA 98:15149-15154

11.47 Rajagopalan, D. (2003): A comparison of statistical methods for analysis of high density oligonucleotide array data. Bioinformatics, **19**, 1469-1476

11.48 Ressom H, Wang D, Natarajan P. (2003): Adaptive double self-organizing maps for clustering gene expression profiles. Neural Netw. **16**, 633-40

11.49 Ringnér, M., Eden, P., Johansson, P. (2002): Classification of expression patterns using artificial neural networks. P. 201-215. In D.P. Berrar, W. Dubitzky, and M. Granzow (Eds.). A Practical approach to microarray data analysis. Kluwer Academic Publishers, Boston

11.50 Ringnér, M., Peterson, C. (2003): Microarray-based cancer diagnosis with artificial neural networks. BioTechniques. **34**, S30-S35

11.51 Ross, D. T., Scherf, U., Eisen, M. B., Perou, C. M., Rees, C., Spellman, P., Iyer, V., Jeffrey, S. S., Van de Rijn, M., Waltham, M. et al. (2000): Systematic variation in gene expression patterns in human cancer cell lines. Nat. Genet. **24**, 227-235

11.52 Russon, E. (2003): Chip critics countered. The Scientist. **17**, 30-31

11.53 Shipp, M., Ross, K., Tamayo, P., Weng, A., Kutok, J., Aguiar, R., Gaasenbeeck, M., Angelo, M., Reich, M. Pinkus, G. et al. (2002): Diffuse large B-cell lymphoma outcome prediction by gene-expression profiling and supervised machine learning. Nat. Med. **8**, 68-74

11.54 Toronen, P., Kolehmainen M, Wong G, Castren E. (1999): Analysis of gene expression data using self-organizing maps. FEBS Lett. **451**, 142-6

11.55 Tseng, G. C., Oh, M. K., Rohlin, L., Liao, J. C., Wong, W. H. (2001): Issues in cDNA microarray analysis: quality filtering, channel normalization, models of variations and assessment of gene effects. Nuc. Acids Res. **29**, 2549-2557

11.56 Werbos, P. (1974): Beyond regression: new tools for prediction and analysis in the behavioral sciences. Doctoral dissertation. Harvard, Cambridge, Mass

11.57 Valafar, F. (2002): Pattern recognition techniques in microarray data analysis: a survey. Ann N Y Acad Sci. **980**, 41-64

12. An Evolutionary Modularized Data Mining Mechanism for Financial Distress Forecasts

Po-Chang (P.C.) Ko[1] and Ping-Chen (P.C.) Lin[2]

[1] National Kaohsiung University of Applied Sciences Department of Information Management Kaohsiung, Taiwan 807, R.O.C
cobol@cc.kuas.edu.tw
[2] Van Nung Institute of Technology Department of Information Management Jungli, Taiwan 320, R.O.C
lety@cc.vit.edu.tw

Abstract: More precise forecasting of corporate financial distress provides important judgment principles to decision-makers, such as bank loan officers, creditors, stockholders, bondholders, government officials and even general public. In this article, we introduce a modularized financial distress forecasting mechanism based on evolutionary algorithm, which allows using any evolutionary algorithm, such as Particle Swarm Optimization, Genetic Algorithm and etc., to extract the essential financial patterns. One more evaluation function modules, such as Logistic Regression, Discriminant Analysis, Neural Network, are integrated to obtain better forecasting accuracy by assigning distinct weights, respectively. For eliminating unreasonable results among these modules, the rule-based evaluation criteria are designed in our mechanism. From our experiments, applying evolutionary algorithm to select critical financial ratios obtains better forecasting accuracy, and, a much better accuracy is obtained if more function modules are integrated in our mechanism.

12.1 Introduction

The corporate financial distress forecasting plays an increasing important role, because the world economy changes radically. It will help many professionals, such as bank loan officers, creditors, stockholders, bondholders, financial analysts, governmental officials and even general public, to make more precise decision in current intense commercial competition environment. But, it is very challenging to extract/discover valuable information from ubiquitous financial statements in various homogeneous/heterogeneous databases.

Due to the radical changes of corporate finance and world economic environment, the critical financial patterns change dynamically [12.1]. Conventionally, the Factor Analysis [12.2]-[12.4] and Stepwise method [12.5] -[12.7] are commonly used to discover the critical financial patterns. But, the discovering financial patterns and forecasting accuracy depend on specific selection sequence deeply, because of their linear searching characteristic. The Genetic

Algorithms *(GA)* [12.8], Evolutionary Programming *(EP)* [12.9], Evolutionary Strategies *(ES)* [12.10] and Genetic Programming *(GP)* [12.11], are four well-known evolution based algorithms. Among them, *GA* is proven to be suitable for feature selection problems [12.12]-[12.16]. Like other evolutionary techniques, the Particle Swarm Optimization *(PSO)* is a newer evolutionary technique, which was first introduced by Eberhart and Kennedy [12.17]-[12.19]. The discrete binary version of *PSO* performs well when solving the problems in a space featuring discrete and qualitative distinctions between variables [12.20]. From our past research, *PSO* achieves better performances than *GA* [12.21].

Currently, the Data Mining or Knowledge Discovery *(KDD)* technologies are rapidly developed to discover interesting patterns in great amount of data, especially in relative financial domain problems. The growing efforts are made to combine machine learning and statistical methodologies in academic and industry researches so far. From the literature reviews, applying machine learning algorithms, such as Neural Networks *(NNs)*, Rule Induction *(RI)* and Fuzzy Logic, achieves better forecasting accuracy than statistics methodologies, such as Discriminant Analysis *(DA)* and Logistic Regression *(LR)* in predicting bankruptcy of corporations [12.22]-[12.24]. In addition, combining more methodologies also further enhances its prediction accuracy than using single methodology [12.5], [12.24]-[12.28].

In this article, we proposed a mechanism to combine both the statistical methods and machine learning algorithms effectively. First, our mechanism extracts critical financial patterns based on one evolutionary algorithm. The various predicting methods are modularized into function modules and integrated by assigning distinct weights. It is noted that each assigned weight is not a fixed value, but learned from our mechanism. For eliminating unreasonable results among these modules, the rule-based evaluation criteria are designed in our mechanism. The financial pattern extraction is evaluated by the integrated predicting method and the rule-based evaluation criteria. Finally, the best financial patterns with highest evaluation value are obtained.

The rest of this article is organized as follows. Our mechanism and the functionalities of each process are proposed in Section 12.2. To prove its feasibility, an implementation is introduced in Section 12.3. In Section 12.4, our results are compared with other methodologies. Finally, conclusions are made to our proposed mechanism about the financial distress forecasting.

12.2 Evolutionary Modularized Mining Mechanism

An evolutionary modularized data mining mechanism shown in Fig. 12.1 is proposed to forecast financial distress from financial statements, which includes three processes: Exploration Process, Transformation Process and Mining Process. The Exploration Process extracts effective data from financial statements stored in various homogeneous/heterogeneous databases into

Data Warehouse. The Transformation Process consolidates these financial data into forms appropriate for Mining Process. Finally, the Mining Process is further divided into two processes: Evolutionary Extraction Process *(EEP)* and Modularized Evaluation Process *(MEP)*. It will discover critical financial patterns dominating corporate distress. Each process is described in the following sections.

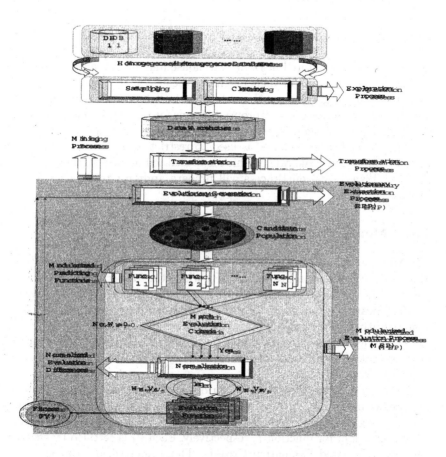

Fig. 12.1. The evolutionary modularized mining mechanism

12.2.1 Exploration Process

Initially, all of the financial statements are raw data in the Exploration Process, which may be widespread in various homogeneous/heterogeneous databases. First, the Sampling Task delimits the analysis categories, which is sufficiently representative of the whole, rather than entire database to reduce processing complexity. Then, the Cleaning Task fixes or discards those raw data sources with flaws, because the databases are highly susceptible to noisy, missing and inconsistent data in real world. Filling constant value (e.g. -999.99) and eliminating the missing data directly are regularly used techniques in Cleaning Task. Finally, the correct and clean sample data is stored in the Data Warehouse.

12.2.2 Transformation Process

In general, the data stored in Data Warehouse may be not suitable for Data Mining Process because of their magnitude scales, analysis units and so on. The Smoothing, Aggregation, Normalization and etc. methods are commonly used to transform them into appropriate forms for mining. In our mechanism, the Aggregation method is used to aggregate daily data into monthly data and the Normalization method is used to restrict each financial magnitude within a particular scale range, which is illustrated in next section. It means that the Transformation Process will prepare appropriate data for Mining Process.

12.2.3 Mining Process

The Mining Process consists of Evolutionary Extraction Process *(EEP)* and Modularized Evaluation Process *(MEP)*, where the former extracts candidate patterns and the latter evaluates its fitness. The Genetic Algorithm *(GA)* and Particle Swarm Optimization *(PSO)* are two famous population-based selection algorithms. First, the financial statements are encoded in *EEP*, where the encoding formats are illustrated in Section 12.3. Then, *EEP* produces next-generation Candidate patterns in each cycle based on the fitness value evaluated from *MEP*.

For compensating the defects of various predicting functions and enhancing predicting accuracy, they are modularized and integrated by assigning distinct weights in *MEP* Process. Let V_f and W_f denote the distinct fitness value and weight for function f, respectively. Each V_f is certified by carefully designed rule-based Evaluation Criteria. Those uncertified $V_f(s)$ are eliminated by assigning $W_f(s)$ to 0. In addition, to diminish judgment complexity for decision makers, our mechanism minimizes the number of extracted patterns denoted as *len*. The integrated fitness value (FV) is function of V_f, W_f and *len* shown in Eq. (12.1), where denotes the evaluation function.

$$FV = \Psi(V_i, W_i, len). \tag{12.1}$$

12.3 Implementations

To prove the feasibility of our mechanism, an implementation is introduced in this section. First, the Sampling and Cleaning Processes eliminates those companies without sufficient effective financial statements since 1991. The *z-score* normalization method, shown in Eq. (12.2), is used to normalize the magnitude of financial ratios, because both the maximum and minimum of γ are unknown, where γ denotes the observation value in financial statement, and γ' is the normalization of γ based on its mean - μ_γ and standard deviation -σ_γ.

$$\gamma' = \frac{\gamma - \mu_\gamma}{\sigma_\gamma}. \qquad (12.2)$$

The *PSO* algorithm is an adaptive algorithm based on the swarm cooperation and competition in the social-psychological science. It performs efficient non-linear search in various problem space. Each particle (X_i) in the candidate patterns is encoded into *(D+1)*-dimensional space, where each dimension is one binary bit in former *D*-dimension, shown in Fig. 12.2. Each bit represents the presence of the corresponding financial ratio, such as R308 (Book Value Per Share), R408 (Total Growth Rate) and etc., shown in Table 12.1. '1' and '0' mean the selected and non-selected state for the corresponding financial ratio, respectively. The latest dimension (θ) is a real number representing the allowable tolerance relative to forecasting accuracy, which is used and further explained in *MEP*.

$$X_i \Rightarrow \boxed{\text{R100} \mid \text{R101} \mid \text{R102} \mid \ldots \mid \text{R836} \mid \theta}$$

Fig. 12.2. Representation of one particle

The particle contains both binary version (B) and normal encoding (N) parts. Let D_B and D_N denote their dimension sizes, respectively. The velocity means the change of probability in former D_B dimension and changing magnitude in latter D_N dimension. Let P_i, PB_i and PG denote the current position of X_i, the best position of X_i so far and the best position of entire population so far, where $P_i = [p_{i1}, p_{i2}, \ldots, p_{i(DB+DN)}]$, $PB_i = [pb_{i1}, pb_{i2}, \ldots, pb_{i(DB+DN)}]$ and $PG = [pg_1, pg_2, \ldots, pg_{(DB+DN)}]$, respectively. The modified *PSO* algorithm is described as follows:

1. Initialize a population of particles with random positions and velocities (v) on $(D_B + D_N)$ dimensions according to its limitation in the problem spaces.
2. For each particle, evaluate its fitness value (V_i) in our *MEP* Process.
3. Let *pbest_i* denote the best solution, which X_i has achieved so far. Compare V_i with *pbest_i*. If (V_i) is better, set *pbest_i* = (V_i) and record current position in PB_i.

Table 12.1. The bit description of particle

1	R100	Return On Total Assets%[1]
2	R101	Return on Total Assets%[2]
3	R102	Return on Total Assets%[3]
4	R103	Return on Equity%[4]
5	R104	Return on Equity%[5]
6	R205	Cash Flow Operating/Current Liabilities%
7	R206	Cost of Interest-Bearing Debt
8	R301	Earning Per Share
9	R302	Book Value Per Share
10	R303	Cash Flow Per Share
11	R304	Sales Per Share
12	R305	Operating Income Per Share
13	R306	Pre-Tax Income Per Share
14	R307	Book Value Per Share
15	R308	Book Value Per Share
16	R501	Current Ratio
17	R502	Acid Test
18	R503	Interest Expense /Total Revenue
19	R504	D/E Ratio
20	R505	Liabilities /Total Assets
21	R506	Equity/Total Assets
22	R507	(Long Term Liabilities+Shareholders' Equity) /Fixed Assets
23	R509	Interest-Bearing Debt/Shareholders' Equity
24	R513	Operating Income Per Shares
25	R514	Pre-Tax Income /Capital
26	R531	Net Income - Exclude Disposal[6]
27	R532	Net Income Ratio-Exclude Disposal[7]
28	R606	(Inventory +Accounts Receivables) /Shareholders' Equity
29	R607	Total Asset Turnover
30	R608	Accounts Receivables Turnover
31	R610	Inventory Turnover
32	R611	Days Inventory Outstanding
33	R612	Fixed Asset Turnover
34	R613	Equity Turnover
35	R614	Days Payables Outstanding
36	R615	Net Operating Cycle
37	R834	Sales Per Employee
38	R835	Operation Income Per Employee
39	R836	Fixed Assets Per Employee

R^*: Gain on Disposal of Investment+Reverse of Loss on Invest Revaluation-Loss on Disposal of Investment-Loss on Invest Revaluation+Gain on Disposal of Fixed Assets-Loss on Disposal of Fix Assets

1: (Pre-Tax Income-(R^*+Interest Expenses +Depreciation+Amortization)*(1-25%))

2: (Ordinary Income+Interest Expenses*(1-25%))/Total Assets

3: (Ordinary Income -R^*)

4: Ordinary Income/Total Stockholders' Equity

5: (Ordinary Income-R^*)/Total Stockholders' Equity

6: Ordinary Income+R^*

7: (R531/Net Sales)*100%

4. Similarly, let *gbest* denote the population's overall previous best. Compare (V_i) with *gbest*. If (V_i) is better, set $gbest_i = (V_i)$ and record the index of P_i.

5. Adjust the velocity and position of X_i according to Eqs. (12.3) and (12.4-12.5), respectively, where c_1 and c_2 are constants, $rand()$ is a random number generator and $S(\cdot)$ is a sigmoid function.

$$v_{id} = v_{id} + c_1 \times rand() \times (pb_{id} - p_{id}) + c_2 \times rand() \times (pg_{id} - p_{id}),$$
$$where \ d = 1, 2, ..., D_B + D_N \qquad (12.3)$$

$$if \ (rand() < S(v_{id})) \ then \ p_{id} = 1 \ else \ p_{id} = 0,$$
$$where \ d = 1, 2, ..., D_B \qquad (12.4)$$

$$p_{id} = p_{id} + v_{id}, \ where \ d = D_B + 1, D_B + 2, ..., D_B + D_N \qquad (12.5)$$

6. Loop to step 2, until the termination criterion is met, such as the sufficient good fitness or maximum iteration.

The Logistic Regression (LR), Discriminant Analysis (DA) and Neural Network (NN) are modularized in *MEP*. Let ρ_f denote the raw evaluation value by function f, $f \in F$, $F = LR, DA, NN$ is the set of function modules. κ_f denotes the classification threshold for function f, where both κ_{LR} and κ_{NN} are assigned to 0.5, and κ_{DA} are calculated by DA function. Then, $\delta_f = \rho_f - \kappa_f$. If $\delta_f > 0$, it means distress is predicted by function f, otherwise, non-distress is predicted. In our implementation, δ_{LR}, δ_{DA} and δ_{NN} are calculated, independently. Each δ_f is normalized to δ_f' by Eq. (12.6), where $S(\cdot)$ is a sigmoid function. Let ξ_{Real} denote the real financial distress state and ξ_f denote the prediction state by using Function f, shown in Eq. (12.7), where 1 means distress state and -1 means non-distress state. $(\xi_f \cdot \xi_{Real})_c = 1$ means correct prediction for company c, otherwise means error.

$$\delta_f' = S(\delta_f) - 0.5, \ where \ S(\delta_f) = \frac{1}{(1 + exp(-\delta_f))} \qquad (12.6)$$

$$\xi_f = \begin{cases} 1, & if \ \delta_f' > 0 \\ -1, & \text{otherwise.} \end{cases} \qquad (12.7)$$

Let C be the set of our target companies, $\Phi(\cdot)$ be a unity function defined in Eq. (12.8). η_f denotes the forecasting accuracy by using Function f, which is illustrated in Eq. (12.9). η_{MAX} denotes the maximum of η_f, where $f \in F$.

$$\Phi(x) = \begin{cases} 1, & if \ x > 0 \\ 0, & \text{otherwise} \end{cases} \qquad (12.8)$$

$$\eta_f = \frac{\sum_{c \in C} \Phi(\xi_f \cdot \xi_{Real})_c}{No. \ of \ C}. \qquad (12.9)$$

For avoiding larger δ'_f and minimum η_f to mislead the forecasting direction in training phase, the allowable tolerance (θ) relative to η_{MAX} is considered. It means that the Evaluation Criterion only considers η_f within $\eta_{MAX} - \theta$. The corresponding weight (W_f) is defined in Eq. (12.10). The integrated forecasting offset δ_F, ξ_F and η_F are calculated from Eqs. (12.11, 12.7) and (12.9), respectively. Finally, the fitness value FV is defined in Eq. (12.12).

$$W_f = \begin{cases} \eta_f, if \ \eta_f < (\eta_{MAX} - \theta) \\ 0, \ \ otherwise \end{cases} \tag{12.10}$$

$$\delta_F = \sum_{f \in F} W_f \cdot \delta'_f \tag{12.11}$$

$$FV = \Psi(\cdot) = \frac{-log_2(1 - \eta_F)}{(1.025)^{len-3}}. \tag{12.12}$$

12.4 Performance Evaluation

Two classes of experiments are designed in this section. One uses single predicting function, such as *LR*, *DA* or *NN*, and the other combines two or more function modules, in *MEP* Process. With assistance of evolutionary algorithm, both of them are capable to extract critical financial patterns. The financial statements of companies are from Taiwan Economic Journal Data Bank *(TEJ)*. All of the simulation programs (*GA, PSO, NN, LR* and *DA*) are written in *Borland C++ Builder 6.0*. The Stepwise statistical method was simulated in *SPSS 8.0* package. All of them were run on the *Microsoft Windows 2000 Professional* platform.

Table 12.2. Distribution of target companies

	Distressed set	Non-distressed set	Total
Training set	87	265	352
Test set	45	140	185
Total	132	405	537

12.4.1 Exploration Process

There are 663 and 425 companies are listed in *Taiwan Stock Exchange (TSE)* and *Gretai Securities Market* until April 2003. 39 financial ratios and 537 companies shown in Table 12.1 and Table 12.2 are selected to be our analysis targets, if eliminating those companies without sufficient effective financial values since 1991 from Exploration Process. These selected companies are classified into distressed/non-distressed set and training/test set, which are shown in Table 12.2. The ratio of distressed to non-distressed companies is 1:3, approximately, because it provides better verification results [12.29]-[12.30]. They are also divided into training set and test set, whose distribution ratio is near to 2:1 [12.31]. Most of these target companies belong to small and middle enterprize, because their net sales and total assets are less than mean value ($81.42 and $622.12 million), respectively. Besides, most of them attract great investors, because their mean market value of equity $452.98 is significantly larger than net sales, the mean leverage (0.62) falls into normal range, and the current ratio (0.66) is lower than normal range. Although these target companies create greater profits, they appear higher risk of daily turnover because their mean are 0.45 and 7.81, respectively. All of these descriptive statistics are summarized in Table 12.3.

Table 12.3. Descriptive statistics for our target companies

	Mean	No.>Mean	Max.	Min.
Net sales (millions)	81.42	194	565.31	1.23
Total asset (millions)	622.12	201	3969.21	34.2
Market value of equity (millions)	452.98	154	4531.11	9.1
Return on total assets (%)	0.45	214	26.19	-47.53
Current ratio (%)	0.66	172	4.96	0.14
Leverage (%)	0.62	198	1.05	0.33
Turnover (%)	7.81	173	12.27	0.01

1. The variables of 537 sample companies are calculated from the TEJ in 2002 September unless otherwise specified.
2. Market value of equity: closing stock price in 12/30/2002 e the number of shares outstanding in 12/30/2002.
3. Return on total assets: EBILAT (Earnings Before Interest, Less Adjusted Taxes) /average total assets.
4. Current ratio: current assets/current liabilities.
5. Leverage: total debt/total assets.
6. Turnover: mean daily trading volume to shares outstanding in 2002.

12.4.2 Mining Process

The parameters of evolutionary algorithms *(GA, PSO)* are illustrated in Table 12.4. To emphasize the capability of integrating various function modules in our proposed mechanism and implementations, 14 experiments are designed to integrate *LR, DA* and *NN* functions. They are *GA-LR/PSO-LR, GA-DA/PSO-DA, GA-NN/PSO-NN, GA-LRDA/PSO-LRDA,*

GA-LRNN/PSO-LRNN, *GA-DANN/PSO-DANN* and *GA-LRDANN/PSO-LRDANN*, shown in Table 12.5.

Table 12.4. Parameters in PSO and GA

PSO	Value	GA	Value
Population Size	100	Population Size	100
Dimension Size (D_B)	39	Chromosome Size	39
Dimension Size (D_N)	1		
No. of Generation (D_N)	5000	No. of Generation	5000
Maximum Velocity (VD_{max})	6	Selection Method	Integral Roulette
Maximum Velocity (VN_{max})	6	Crossover Method	Two Point
Acceleration Constant (c_1)	2	Crossover Rate	0.7
Acceleration Constant (c_2)	2	Mutation Rate	0.005

Table 12.5. The relations between methods and our experiments

Experiments	EEP	MEP
GA-LR/PSO-LR	GA/PSO	LR
GA-DA/PSO-DA	GA/PSO	DA
GA-NN/PSO-NN	GA/PSO	NN
GA-LRDA/PSO-LRDA	GA/PSO	LR + DA
GA-LRNN/PSO-LRNN	GA/PSO	LR + NN
GA-DANN/PSO-DANN	GA/PSO	DA + NN
GA-LRDANN/PSO-LRDANN	GA/PSO	LR + DA + NN

12.4.3 Simulation Results

To manifest the superiority of our approach, we compare our results with Logistic Regression *(S-LR)*, Discriminant Analysis *(S-DA)* and Neural Network *(S-NN)* based on Stepwise method. The forecasting accuracy of each quarter before distress in Test Phase and their mean value are shown in Table 12.6. From Fig. 12.3, it is worth noting that using evolutionary algorithm *(PSO, GA)* to select critical financial ratios will produce better forecasting accuracy than using conventional statistical method (Stepwise). For instance, $E(\eta_{PSO-DA})$ (0.75) $> E(\eta_{GA-DA})$ (0.71) $> E(\eta_{S-DA})$ (0.61), $E(\eta_{PSO-LR})$ (0.85) $> E(\eta_{GA-LR})$ (0.80) $> E(\eta_{S-LR})$ (0.72) and $E(\eta_{PSO-NN})$ (0.87) $> E(\eta_{GA-NN})$ (0.82) $> E(\eta_{S-NN})$ (0.77), where $E(\eta_f)$ denotes the mean value of η_f. From Fig. 12.4, it also demonstrates that integrating more modularized functions achieves better forecasting accuracy, no matter which evolutionary

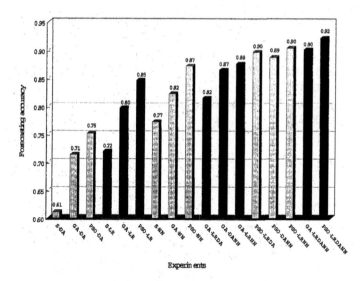

Fig. 12.3. The forecasting accuracies if using PSO and GA selection method over DA, LR, NN, LRDA, LRNN, DANN and LRDANN evaluation function, respectively

Table 12.6. The mean forecasting accuracy of each model (Test Phase)

Experiment	Q_1	Q_2	Q_3	Q_4	Q_5	Q_6	Q_7	Q_8	Mean
S-DA	0.58	0.68	0.6	0.65	0.66	0.60	0.61	0.51	0.61
GA-DA	0.72	0.75	0.74	0.75	0.72	0.68	0.69	0.66	0.71
PSO-DA	0.76	0.76	0.75	0.76	0.74	0.74	0.74	0.74	0.75
S-LR	0.78	0.76	0.75	0.72	0.74	0.68	0.72	0.61	0.72
GA-LR	0.81	0.83	0.84	0.85	0.78	0.79	0.77	0.71	0.80
PSO-LR	0.87	0.88	0.87	0.87	0.84	0.82	0.80	0.82	0.85
S-NN	0.75	0.76	0.74	0.78	0.79	0.81	0.79	0.76	0.77
GA-NN	0.83	0.84	0.86	0.83	0.84	0.82	0.79	0.77	0.82
PSO-NN	0.89	0.92	0.91	0.93	0.82	0.84	0.85	0.82	0.87
GA-LRNN	0.86	0.88	0.90	0.92	0.91	0.89	0.85	0.80	0.88
GA-DANN	0.86	0.90	0.88	0.89	0.86	0.85	0.85	0.83	0.87
GA-LRDA	0.78	0.85	0.86	0.85	0.82	0.79	0.79	0.78	0.82
PSO-LRNN	0.91	0.92	0.91	0.95	0.91	0.84	0.92	0.89	0.90
PSO-DANN	0.89	0.89	0.92	0.93	0.91	0.86	0.86	0.84	0.89
PSO-LRDA	0.89	0.89	0.89	0.93	0.92	0.85	0.92	0.88	0.90
GA-LRDANN	0.93	0.91	0.92	0.94	0.90	0.88	0.87	0.86	0.90
PSO-LRDANN	0.91	0.92	0.94	0.95	0.92	0.92	0.92	0.91	0.92

algorithm is used. For example, Fig 12.4(a) shows that $E(\eta_{PSO-LRDANN})$ (0.92) > $[E(\eta_{PSO-LRDA})$ (0.90), $E(\eta_{PSO-DANN})$ (0.89), $E(\eta_{PSO-LRNN})$ (0.90)], $E(\eta_{PSO-LRDA})$ (0.90) > $[E(\eta_{PSO-LR})$ (0.85), $E(\eta_{PSO-DA})$ (0.75)], $E(\eta_{PSO-DANN})$ (0.89) > $[E(\eta_{PSO-DA})$ (0.75), $E(\eta_{PSO-NN})$ (0.87)], and $E(\eta_{PSO-LRNN})$ (0.90) > $[E(\eta_{PSO-LR})$ (0.85), $E(\eta_{PSO-NN})$ (0.87)], and Fig. 12.4(b) shows that $E(\eta_{GA-LRDANN})$ (0.90) > $[E(\eta_{GA-LRDA})$ (0.82), $E(\eta_{GA-DANN})$ (0.87), $E(\eta_{GA-LRNN})$ (0.88)], $E(\eta_{GA} - LRDA)$ (0.82) > $[E(\eta_{GA-LR})$ (0.80), $E(\eta_{GA-DA})$ (0.71)], $E(\eta_{GA-DANN})$ (0.87) > $[E(\eta_{GA-DA})$ (0.71), $E(\eta_{GA-NN})$ (0.82)] and $E(\eta_{GA-LRNN})$ (0.88) > $[E(\eta_{GA-LR})$ (0.80), $E(\eta_{GA-NN})$ (0.82)].

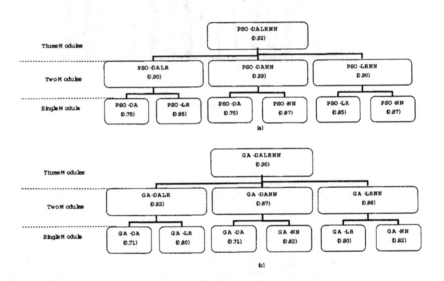

Fig. 12.4. The comparison of the using PSO and GA selection method between single and hybrid modules

Besides, our approach extracts less critical financial ratios to forecast financial distress without losing accuracy. Let NUM_{exp} denote the number of selected financial ratios if applying exp experiment. From Table 12.7, it is obviously that the number of extracted financial ratios by Stepwise is 8 approximately [$NUM_{S-DA} = 8.33$, $NUM_{S-LR} = 7.92$, $NUM_{S-NN} = 8.02$], but, using evolutionary algorithms only extract 3 to 5 financial ratios [$NUM_{GA-DA} = 4.22$, $NUM_{GA-LR} = 3.38$, $NUM_{GA-NN} = 5.02$, $NUM_{GA-LRNN} = 3.92$, $NUM_{GA-LRDA} = 4.42$, $NUM_{GA-DANN} = 4.82$, $NUM_{GA-LRDANN} = 4.93$ (GA Algorithm), $NUM_{PSO-DA} = 4.14$, $NUM_{PSO-LR} = 3.71$,

$NUM_{PSO-NN} = 4.58$, $NUM_{PSO-LRNN} = 4.07$, $NUM_{PSO-LRDA} = 4.11$, $NUM_{PSO-DANN} = 4.32$, $NUM_{PSO-LRDANN} = 4.93$ (PSO Algorithm)].

Table 12.7. The number of financial ratios selected from various experiment

Experiment	NUM	Experiment	NUM	Experiment	NUM
S-DA	8.33	GA-DA	4.22	PSO-DA	4.14
S-LR	7.92	GA-LR	3.38	PSO-LR	3.71
S-NN	8.02	GA-NN	5.02	PSO-NN	4.58
		GA-LRNN	3.92	PSO-LRNN	4.07
		GA-LRDA	4.42	PSO-LRDA	4.11
		GA-DANN	4.82	PSO-DANN	4.32
		GA-LRDANN	4.93	PSO-LRDANN	4.93

12.5 Conclusions

Due to the radical change in world economic environment, more precise forecasting of corporate financial distress provides important judgment principle to decision-makers. It attracts more interests to integrate the machine learning technologies and conventional statistical tools for solving financial related problems, so far. Especially, the evolutionary algorithms achieve efficient optimization in the problem domain with nonlinear searching capability. In this article, we introduce a financial distress forecasting approach based on evolutionary algorithms, such as *GA, PSO* and etc. From our implementations and experiments, the evolutionary algorithms obtain better forecasting accuracy with minimum critical financial patterns than Stepwise methods (conventional statistical techniques). In addition, we propose a modularized approach to integrate more function modules by assigning distinct weights for achieving better predicting accuracies. It is noted that each assigned weight is not a fixed value, but learned from our mechanism. Our experiments also show that more function modules integrated acquires more accurate forecast. The evaluation criteria design plays an indispensable role in our mechanism. From our simulation results, carefully designing the criteria effectively eliminates those unreasonable results produced by some function modules and helps further improving forecasting accuracy; otherwise, exactly opposite effect is achieved The financial pattern extraction is evaluated by the integrated predicting method and the rule-based evaluation criteria. Besides, a modified *PSO* algorithm is proposed allowing encoding binary and real data within one particle simultaneously. Finally, the most critical financial patterns with highest predicting accuracy are obtained.

References

12.1 John, S. G., Robert, W. I. (2001): Tests of the generalizability of altman's bankruptcy prediction model. Journal of Business Research, **54**, 53–61

12.2 Altman, E. I. (1968): Financial ratios discriminant analysis and the prediction of corporate bankruptcy. The Journal of Finance, **4**, 589–609

12.3 Blum, M. (1974): Failing company discriminant analysis. Journal of Accounting Research, 1–25

12.4 Deakin, E. B. (1972): A discriminant analysis of predictors of business failure. Journal of Accounting Research, 167–179

12.5 Jo, H., Han, I. (1996): Integration of case-based forecasting, neural network, and discriminant analysis for bankruptcy prediction. Expert Systems With Applications, **4**, 415–422

12.6 Laitinen, E. K., Laitinen, T. (2000): Bankruptcy prediction application of the Taylors expansion in logistic regression. International Review of Financial Analysis, **9**, 327–349

12.7 Shirata, C. Y. (2000): Peculiar behavior of Japanese bankrupt firms: Discovered by AI-based data mining technique. In Fourth International Conference on Knowledge-Based Intelligent Engineering Systems & Allied Technologies, 663–666

12.8 Holland, J. (1975): Adaptation in natural and artificial systems: an introductory analysis with applications to biology, control and artificial intelligence. MIT Press

12.9 Fogel, L. J. (1994): Evolutionary programming in perspective: the top-down view. in Computational Intelligence: Imitating Life, Zurada, J. M., Marks, R. J., Robinson, C. J., Eds., IEEE Press

12.10 Rechenberg, I. (1994): Evolutionary strategy. in Computational Intelligence: Imitating Life, Zurada, J. M., Marks, R. J., Robinson, C. J., Eds., IEEE Press

12.11 Koza, J. R. (1992): Genetic programming: on the programming of computers by means of natural selection. MIT Press

12.12 Aluallim, H., Dietterich, T. G.. (1992): Efficient algorithms for identifying relevant features. proc. of Canadian Conference on Artificial Intelligence, 38–45

12.13 Jain, A., Zongker, D. (1997): Feature selection: evaluation, application, and small sample performance. IEEE Transaction on Pattern Analysis and Machine Intelligence, 153–158

12.14 Liu H., Setiono, R. (1996): A probabilistic approach to feature selection: a filter solution. Proc. of 13th International Conference on Machine Learning, 319–327

12.15 Emmanouilidis, C., Hunter, A., MacIntyre, C. C. (1999): Multiple-criteria genetic algorithms for feature selection in neurofuzzy modeling. International Joint Conf. on Neural Networks, 4387–4392

12.16 Srinivas, M., Lalit, M. P. (1994): Genetic algorithms a survey. IEEE Computer, 18–20

12.17 Eberhart, R. C., Kennedy, J. (1995): A new optimizer using particle swarm theory. Proc. 6th International Symposium on Micro Machine and Human Science, Nagoya, Japan, IEEE Service Center, 39–43

12.18 Kennedy, J., Eberhart, R. C. (1995): Particle swarm optimization. Proc. of IEEE International Conf. on Neural Networks, Perth, Australia, IEEE Service Center, 1942–1948

12.19 Kennedy, J. (1997): The particle swarm: social adaptation of knowledge. Proc. of IEEE International Conf. on Evolutionary Computation, Indianapolis, Indiana, IEEE Service Center, 303–308

12.20 Kennedy, J., Eberhart R. C. (1997): A discrete binary version of the particle swarm algorithm. Proc. of International Conf. on Systems, Man, and Cybernetics, IEEE Service Center

12.21 Ko, P. C., Lin, P. C. (2003): A hybrid swarm intelligence based mechanism for earning forecast. accepted by Asian Journal of Information Technology

12.22 Wilson, R. L., Sharda, R. (1994): Bankruptcy prediction using neural networks. Decision Support Systems, 11, 545–557

12.23 Sung, T. K., Chang, N., Lee, G. (1999): Dynamics of modeling in data mining: interpretive approach to bankruptcy prediction. Journal of Management Information Systems, 16, 63–85

12.24 Wallrafen, J., Protzel, P., Popp, H. (1996): Genetically optimized neural network classifiers for bankruptcy prediction. In Proceedings of the 29th Annual Hawaii International Conference of System Sciences, 419–426

12.25 Lin, F. Y., McClean, S. (2001): A data mining approach to the prediction of corporate failure. Knowledge-Based System, 14, pp.189–195

12.26 Ahn, B. S., Cho, S. S., Kim, C. Y. (2000): The integrated methodology of rough set theory and artificial neural network for business failure prediction. Expert Systems With Applications, 18, 65–74

12.27 Lee, K. C., Han, I., Kwon, Y. (1996): Hybrid neural network models for bankruptcy predictions. Decision Support Systems, 18, 63–72

12.28 Donate, J. M., Schryver, J. C., Hinkel, C. C., Schmoyer, R. L., Leuze, M. R., Grandy, N. W. (1999): Mining multi-dimensional data for decision support. Future Generation Computer Systems, 15, 433–441

12.29 Ohlson, J. A. (1980): Forecasting financial failure a re-examination. Journal of Accounting Research, 109–131

12.30 Zmijewski, M. E. (1984): Methodological issues related to the estimation of financial distress prediction model. Journal of Accounting Research, 59–82

12.31 Haykin, S. (1999): Neural networks a comprehensive foundation. Prentice Hall

Subject Index

Printing: Strauss GmbH, Mörlenbach
Binding: Schäffer, Grünstadt